Navigating VMware Turmoil in the Broadcom Era

Insights and Strategies for Transitioning to Alternative Solutions

Sumit Bhatia
Chetan Gabhane

Apress®

Navigating VMware Turmoil in the Broadcom Era: Insights and Strategies for Transitioning to Alternative Solutions

Sumit Bhatia
Houston, TX, USA

Chetan Gabhane
Pune, Maharashtra, India

ISBN-13 (pbk): 979-8-8688-1263-7
https://doi.org/10.1007/979-8-8688-1264-4

ISBN-13 (electronic): 979-8-8688-1264-4

Managing Director, Apress Media LLC: Welmoed Spahr
Acquisitions Editor: Aditee Mirashi
Development Editor: James Markham
Editorial Assistant: Kripa Joseph

Cover designed by eStudioCalamar

Cover image designed by Pixabay

Distributed to the book trade worldwide by Springer Science+Business Media New York, 1 New York Plaza, Suite 4600, New York, NY 10004-1562, USA. Phone 1-800-SPRINGER, fax (201) 348-4505, e-mail orders-ny@springer-sbm.com, or visit www.springeronline.com. Apress Media, LLC is a California LLC and the sole member (owner) is Springer Science + Business Media Finance Inc (SSBM Finance Inc). SSBM Finance Inc is a **Delaware** corporation.

For information on translations, please e-mail booktranslations@springernature.com; for reprint, paperback, or audio rights, please e-mail bookpermissions@springernature.com.

Apress titles may be purchased in bulk for academic, corporate, or promotional use. eBook versions and licenses are also available for most titles. For more information, reference our Print and eBook Bulk Sales web page at http://www.apress.com/bulk-sales.

Any source code or other supplementary material referenced by the author in this book is available to readers on GitHub. For more detailed information, please visit https://www.apress.com/gp/services/source-code.

If disposing of this product, please recycle the paper

To the trailblazers of the virtualization landscape.

This book is dedicated to the IT leaders, engineers, and visionaries who confront uncertainty with bravery, transforming obstacles into opportunities for innovation. In the challenging landscape shaped by Broadcom and VMware, your resilience and creativity motivate us to trust in the transformative potential of technology and the human spirit.

As co-authors, we, Sumit and Chetan, have infused our shared enthusiasm, extensive research, and numerous discussions into these pages. This journey was not merely about writing a book; it was about engaging with a community that shares our commitment to progress, adaptability, and the bravery to explore new avenues.

We aspire for this guide to serve as more than just a navigation tool for the virtualization realm. May it stand as a tribute to perseverance, a source of clarity during uncertain times, and a catalyst for innovation that empowers you to create a future filled with possibilities.

Table of Contents

About the Authors

 **Sumit Bhatia**, the lead author, is a distinguished global infrastructure solutions architect dedicated to tackling complex business challenges while promoting Green IT through innovative hybrid cloud strategies, advanced automation, and cost efficiency in IT operations to support the global oil and gas sector. With over 16 years of experience in the technology industry, Sumit is at the forefront of technological advancements, leveraging his extensive expertise in automation techniques, DevOps methodologies, multi-cloud environments, and edge computing solutions engineering. He has gained recognition in the industry for his numerous contributions, including his authorship of the acclaimed book *Reverse Engineering with Terraform*, which has garnered praise from numerous industry experts.

Through his groundbreaking work with Terraform, the implementation of capacity management as a service (CMaaS) in hybrid environments, and the introduction of innovative ON/OFF solutions, his organization achieved significant cost savings, reducing annual expenditures by hundreds of thousands of dollars. He has been honored with the "highest honor" award from his organization, a leader in energy innovation. Additionally, Sumit is a prominent author in leading journals within the oil and gas industry and actively engages with the community as an expert reviewer for top-tier technology publications and journals. His unwavering commitment to sharing knowledge and best practices has greatly assisted many professionals in navigating the intricate landscape of sustainability and solutions engineering.

Chetan Gabhane is a highly experienced professional with over 16 years in the fields of solution architecture and senior technical consulting, specializing in hybrid and multi-cloud computing. His career is marked by a strong commitment to innovation, helping organizations effectively integrate on-premises systems with cloud technologies through the development of robust, scalable, and secure hybrid cloud solutions.

Chetan is well regarded for his deep expertise in cloud ecosystems, where he excels at creating solutions that balance agility with operational efficiency. He focuses on utilizing advanced DevOps practices to enhance development processes and optimize cloud environments for both performance and cost savings. His technical skills also encompass the implementation of automation frameworks that facilitate seamless integrations, enabling businesses to reach their digital transformation objectives.

As the author of the widely recognized book *Reverse Engineering with Terraform*, Chetan has positioned himself as a thought leader in the cloud computing sector. This work, praised for its thoroughness and practicality, has assisted numerous professionals in mastering Terraform and simplifying complex cloud infrastructures.

About the Technical Reviewer

 **Christophe Lombard** is an IT architect with 28 years of experience in designing and delivering complex solutions in both consultative and technical leadership positions with a specific focus on Cloud, IT transformation and Cybersecurity. He has worked with large organizations like NEC, CSC, DELL EMC, and more recently at VMware. He has helped dozens of IT professionals and organizations achieve their business objectives through business and consultative engagements. During his career, he has served as a network and security engineer, project manager, consultant, and cloud architect. He started developing his knowledge in VMware in 2005 and his cloud expertise in 2015 and has developed his security knowledge all over his career. He is passionate about the development of innovation in companies using new technologies: cloud, cybersecurity, microservices, big data, and artificial intelligence.

His multiple experiences and areas of expertise opened a door for him at VMware in 2020 during the pandemic and very recently at Palo Alto Networks. As a Cloud Security Solution Architect, Christophe helps drive Prisma Cloud product adoption within large enterprises and customers. He loves to learn, to enable, and to educate people, including customers, partners, and colleagues, on all the cloud and cybersecurity technologies he is focused on. Christophe holds an AWS Certified Solution Architect-Associate certification, a GCP Associate Cloud Engineer certification, and a HashiCorp Certified: Terraform Associate certification.

In his spare time, he enjoys working on his creative pursuits such as photography. Find Christophe at linkedin.com/in/lombardchristophe.

Acknowledgments

The process of writing this book has been an extraordinary experience, one marked with challenges, late nights, extensive discussions, and a collective dedication to create something impactful. This work represents not merely a book but the synthesis of our shared knowledge, experiences, and a profound aspiration to assist others in navigating unfamiliar territories.

We extend our deepest gratitude to our families. Your patience, understanding, and steadfast faith in us have been invaluable. You supported us during moments of deep contemplation, surrounded by notes, or when we were engrossed in our screens. Without your love and encouragement, this project would not have come to fruition.

To our mentors and colleagues, we appreciate your role in shaping our comprehension of technology and for motivating us to question, learn, and evolve. Each conversation, debate, and shared insight has contributed significantly to the foundation of knowledge presented in this book.

We are also immensely grateful to the IT community professionals, leaders, and visionaries who have navigated the challenges of the Broadcom-VMware era with resilience and tenacity. Your narratives, challenges, and victories inspired this book. We aimed to honor your experiences by offering practical strategies and actionable insights.

This book required more from us than we anticipated – numerous hours of research, in-depth analysis, and the courage to challenge established norms. It pushed our boundaries but also reinforced our commitment to providing something genuinely valuable to the technology community.

ACKNOWLEDGMENTS

Lastly, to you, the reader, thank you for placing your trust in us. We crafted this book with the hope that it will not only assist you in overcoming the challenges of this new era but also motivate you to embrace its opportunities. Your success is our ultimate reward.

With sincere appreciation,
Sumit and Chetan

Introduction

In an era characterized by digital transformation, virtualization has become a fundamental element of contemporary IT infrastructure. However, as we enter the Broadcom-VMware phase, the journey ahead is laden with challenges, uncertainties, and crucial choices. This book aims to steer you through these challenging times, providing not only strategies but also a clear sense of direction and assurance.

Why This Book?

The acquisition of VMware by Broadcom represents a significant turning point in the IT sector, with extensive consequences for organizations globally. Changes in licensing, rising costs, and the risk of vendor lock-in have prompted many organizations to reevaluate their future strategies. In such a climate, possessing a well-informed roadmap is essential.

What's Inside?

This book is designed to be both thorough and practical. We analyze the transition from Broadcom to VMware and investigate alternative solutions such as Microsoft Hyper-V, Nutanix, Proxmox, and RedHat. We cover the technical, strategic, and financial aspects that can assist you in navigating this transition. Additionally, real-world case studies, best practices, and actionable insights enrich our discussions.

Who Is This Book For?

This book is tailored for CIOs, IT managers, solution architects, and technology enthusiasts alike. It aims to empower readers by providing clarity for technocrats, actionable strategies for decision-makers, and stability for organizations during turbulent times.

Our Commitment to You

As co-authors, we approach this book not only as technologists but as allies in your journey. We have traversed similar uncertain paths, encountered comparable challenges, and adapted to industry changes. This book serves as a platform for sharing our insights, offering guidance, and supporting you as you navigate your own course.

We aspire for this book to equip you with the necessary tools and knowledge to confidently and adeptly navigate the Broadcom-VMware era. Let us embrace this opportunity together to innovate and create a future that's stronger, smarter, and more adaptable.

With sincere dedication,
Sumit and Chetan

CHAPTER 1

The Broadcom Era Begins

Ever since server virtualization technology was introduced to the IT infrastructure landscape, VMware has established itself as a leader in the industry. With a strong following among Fortune organizations globally, VMware has become synonymous with server virtualization. However, the recent acquisition of VMware by Broadcom has sparked uncertainty about the future of this widely used technology. In this chapter, we will take an in-depth look at the VMware acquisition by Broadcom. Additionally, we will analyze previous acquisitions by Broadcom and the industry trends they have set. Lastly, we will assess the potential implications of this acquisition on the IT sector.

1.1 Understanding the VMware Acquisition by Broadcom

The VMware acquisition by Broadcom has sent ripples through the tech industry, raising questions about the future of private cloud environments and the strategic direction of two major players in the field. In this section, we will delve into the details of this acquisition, exploring its implications, potential impacts, and what it means for VMware customers and partners.

© Sumit Bhatia and Chetan Gabhane 2025
S. Bhatia and C. Gabhane, *Navigating VMware Turmoil in the Broadcom Era*,
https://doi.org/10.1007/979-8-8688-1264-4_1

Broadcom's acquisition of VMware for approximately $61 billion caused a stir in the computer industry. This strategic move was aimed at expanding Broadcom's footprint in the corporate software sector. The acquisition not only aligns with Broadcom's goal of becoming a leading infrastructure technology company but also allows for their product line diversification. Following the completion of the acquisition on November 22, 2023, VMware's common stock ceased trading on the New York Stock Exchange (NYSE).

Historically, VMware has had a number of owners. It was previously owned by Dell, which took over the server virtualization company following its $67 billion purchase of EMC in 2016. EMC had purchased VMware in 2004. In 2021, Dell spun out its share of VMware, paving the way for the Broadcom acquisition.

The acquisition of VMware by Broadcom has sparked concerns and had a significant impact on managed service providers (MSPs) and VMware users in the industry. Valued at around $61 billion, the acquisition has brought about various changes and hurdles, especially for the server virtualization industry. One of the primary effects of the acquisition is the shift from VMware's traditional perpetual licensing model to a subscription-based one. This change has left many grappling with price hikes and a decrease in standalone options that were previously available. The pricing adjustments have particularly hit smaller MSPs and organizations specializing in cloud services, with some reporting a tenfold increase in expenses when working with VMware. Additionally, the acquisition has caused potential disruptions and necessitated migrations for MSPs, VMware users, and cloud service providers (CSPs) who can no longer resell or provide support for VMware. As a result, VMware users have had to look for the new potential virtualization environment and seek out alternative solutions to meet their clients' and business requirements.

Overall, the impact of the Broadcom acquisition of VMware on industry has been substantial, driving industry to adapt to the new licensing model and seek alternative solutions to provide value to their businesses and clients. It is crucial for the server virtualization industry to understand and navigate the implications of this acquisition in order to remain competitive and continue delivering quality services to their businesses and customers.

1.1.1 Broadcom's Rationale

Broadcom's acquisition of VMware represents a strategic coup that enhances its standing in the fiercely competitive cloud computing industry. VMware's extensive knowledge in virtualization and private cloud infrastructure perfectly complements Broadcom's current range of networking and semiconductor products. This collaboration is anticipated to open up new opportunities for creativity and market growth, harnessing VMware's cutting-edge technologies in conjunction with Broadcom's current offerings. The incorporation of VMware's products confers a substantial edge, bolstering Broadcom's capacity to address the intricate requirements of contemporary businesses with a more comprehensive and unified set of private cloud solutions.

This strategic move by Broadcom underscores the company's deliberate shift toward a more diversified business model that encompasses both hardware and software sectors. Through the integration of VMware's technology, Broadcom is poised to address the rising demand for seamless, scalable, and secure private cloud computing solutions, positioning itself advantageously in the competitive landscape. The acquisition sets the stage for strategic partnerships that promise to deliver cutting-edge innovations across various industries, driving digital transformation and empowering businesses to thrive in the cloud-centric environment.

Furthermore, the potential motivation for integrating VMware into Broadcom's framework is set to expedite creativity, nurturing the emergence of fresh offerings and solutions that have the potential to reshape norms within the sector. This potential tactical partnership is not solely about expanding the range of products but could be concentrated on establishing a strong presence in private cloud and on-prem virtualization technology that foresees and caters to the changing needs of the digital realm. By acquiring VMware, Broadcom aims to establish itself as a significant contender, ready to shape the course of private cloud computing, highlighting the critical role of VMware in its pursuit of market supremacy and technological superiority.

Broadcom's strategy going forward is to possibly enable enterprise customers to create and modernize their private and hybrid cloud environments. It plans to invest in VMware Cloud Foundation (we will discuss more about this in upcoming chapters), the software stack that serves as the foundation of private and hybrid clouds and is switching away from perpetual software licensing. The product family is now called VMware by Broadcom.

The range of VMware offerings encompasses solutions for enhancing and streamlining private cloud and edge environments, such as VMware Tanzu for expediting application deployment, in addition to application networking (load balancing) and cutting-edge security services. VMware software-defined edge caters to the needs of telecommunication and corporate edge infrastructures.

Broadcom's acquisition of VMware marks a significant change in its strategic focus, expanding beyond its traditional infrastructure and semiconductor products to encompass the rapidly growing cloud and software sector. VMware's well-established presence in the corporate market, along with its cutting-edge technologies such as vSphere and Tanzu, gives Broadcom a direct pathway into this rapidly expanding market. By utilizing its expertise in infrastructure and semiconductor technology, Broadcom intends to improve VMware's offerings and strengthen its position as a leading player in the corporate software industry. This acquisition also enables Broadcom to access the substantial revenue streams generated by ongoing software subscriptions and cloud services, further broadening its business portfolio and ensuring sustained growth.

1.1.2 VMware's Perspective

Broadcom's acquisition of VMware has raised significant concerns regarding potential changes in the current ecosystem, value chain for many customers, company culture, and product roadmap itself. Many industry experts have expressed worries about the impact of the acquisition on VMware's culture of innovation and product development efforts. There are legitimate concerns about whether the acquisition will result in a transformation or weakening of VMware's well-established culture of innovation, customer-centric approach, and employee involvement. Furthermore, there are uncertainties about the future of VMware's product roadmap, particularly in terms of ongoing development, support, and investment in current and upcoming technologies. The technology sector is closely monitoring the situation as it progresses, assessing how Broadcom's ownership will shape VMware's strategic decisions, product lineup, and relationships with partners and customers.

It is imperative for all stakeholders, such as customers and partners, to be kept abreast of any changes, transitions, or shifts that may occur as a result of the acquisition, due to the significant influence it could have on company culture and product roadmap. The maintenance of transparent communication among VMware, Broadcom, and their respective ecosystems will play a vital role in effectively handling any potential alterations and guaranteeing a seamless transition for everyone involved.

1.1.3 The Future of VMware Under Broadcom

Broadcom is looking to boost its product lineup by tapping into VMware's virtualization expertise. By incorporating VMware's software-defined data center (SDDC) tools like vSphere and vSAN, Broadcom aims to offer a robust solution for hybrid and multi-cloud setups. There's also potential for Broadcom to merge VMware's CMP, VMware Cloud Management

Platform, with its own services like Symantec's Integrated Cyber Defense Platform, possibly leading to a unified cloud management system covering infrastructure, security, and applications.

Broadcom's go-to-market strategy is expected to adopt a multi-channel approach to marketing VMware's products and services. The company plans to leverage its existing sales force to target enterprise customers while also expanding VMware's distribution channels to reach a broader market. Moreover, just as Dell previously made investments in developing collaborative products utilizing VMware technology, such as VXRail and VXBlock, Broadcom could potentially allocate resources toward VMware's partner network in order to provide holistic solutions to their clientele. Additionally, Broadcom might consider venturing into emerging markets like embedded systems and edge computing, leveraging VMware's technologies to capitalize on potential growth opportunities.

Broadcom's decision to shift VMware's business model from perpetual software licenses to a subscription-based model is in line with its overarching strategy of transitioning its software business toward recurring revenue streams. This strategic move is expected to enhance revenue predictability, foster customer lock-in, and drive higher profit margins for Broadcom in the long run. The amalgamation of these two entities prompts inquiries into how Broadcom's objectives will influence VMware's trajectory. One potential apprehension is that Broadcom might prioritize hardware sales over software innovation, potentially diverting attention away from enhancing VMware's core virtualization products and customer support. Moreover, Broadcom's track record of cost-cutting and layoffs has sparked concerns regarding potential job cuts and reduced investment in VMware's workforce.

Overall, the outlook for VMware under Broadcom's ownership is ambiguous. Despite the possible drawbacks and obstacles linked to the takeover, there are also prospects for expansion and advancement. The effectiveness of the merger will rely on Broadcom's capacity to address the hurdles and capitalize on the synergies between the two entities.

The manner in which Broadcom harmonizes its emphasis on hardware along with VMware's proficiency in software and customer networks will play a pivotal role in determining VMware's trajectory under its fresh management. It is recommended that VMware users remain abreast of the most recent updates and modifications concerning the VMware-Broadcom acquisition.

1.1.4 Integration Challenges

Despite many months that have passed since the acquisition was announced as completed, there remains a significant amount of ongoing turmoil. The integration of the disparate business models and cultures of the two companies continues to pose a significant challenge in the aftermath of the acquisition. Broadcom and VMware have traditionally operated with distinct approaches to business, culture, and product development. Effectively aligning these differences is essential to ensure a seamless transition and to maximize the potential synergies between the two companies.

The following are the key areas for aligning the business models and cultures of Broadcom and VMware that affect their customers.

1.1.4.1 Clear Communication and Transparency

Effective and transparent communication among the leadership teams, employees, and stakeholders of both organizations is crucial for closing the divide and establishing a mutual comprehension of the vision, objectives, and principles of the merged entity. Currently, Broadcom has struggled to deliver unambiguous and persuasive communication regarding their strategic decisions. According to multiple blog entries, numerous VMware stakeholders have expressed dissatisfaction over the lack of information provided about their yearly renewal fees until shortly before their renewal deadlines. This limited timeframe leaves little opportunity for end users to make informed decisions and respond accordingly. The deficiency in communication and transparency is resulting in challenges and trust issues with the VMware stakeholders.

1.1.4.2 Product and Technology Alignment

Following the acquisition by Broadcom, there is currently a lack of clarity in the market regarding the various offerings and software bundles available for customers. The VMware products offered by Broadcom are opulent and do not align with the needs and desires of customers. Broadcom is attempting to entice customers by bundling all VMware cloud products into a single suite known as VMware Cloud Foundation. This move would surely benefit Broadcom as it aims to establish long-term dependencies on VMware's diverse product portfolio, potentially creating a sustainable revenue stream. However, the bundled offering may be excessive for many industry users who do not require these additional products, thus diminishing the overall appeal of the deal. It is crucial to align the product and technology stacks of VMware products with the requirements and preferences of key stakeholders to ensure they are meeting the business needs of their customers.

1.1.4.3 Customer and Partner Engagement

Implementing a customer-focused strategy and nurturing strong partnerships with collaborators are typically crucial in minimizing potential disruptions and showcasing a dedication to providing value to the market. Nevertheless, the merger has resulted in a severe setback for nearly all VMware partners. Broadcom has terminated the VMware partner program with almost all of its partners. The rationale behind Broadcom's decision is to concentrate on direct sales and streamline its business model. Nonetheless, this action has raised apprehension among VMware's partners, who are uncertain about their future association with the company. The discontinuation of the program affects a broad spectrum of partners, including value-added resellers (VARs), system integrators (SIs), and managed service providers (MSPs). These partners have traditionally depended on the VMware Partner Program for incentives, training, and assistance. The termination has forced them to urgently reevaluate their business strategies.

Industry sources confirm that Broadcom is redoing this partner program only with 100 renowned partners at this stage. Here is the link to the FAQ document released by Broadcom for the partners to transition from the VMware partner program to Broadcom Advantage (`https://docs.broadcom.com/doc/vmware-partner-faq`). The decision has also raised questions about the future of VMware's ecosystem and its commitment to the channel. Some partners fear that Broadcom's focus on direct sales could lead to a decline in the availability of VMware products and services through indirect channels. Additionally, the cancellation of the program could hinder the growth of new and emerging partners who have relied on VMware's support to establish themselves in the market. The impact of this move on VMware's market share and revenue remains to be seen, but Broadcom's decision will have far-reaching consequences for the IT industry.

1.1.4.4 Adapting Business Models

Broadcom's aggressive approach toward VMware products is not proving beneficial in the current scenario. VMware boasts a significant customer base and a solid reputation in the field of virtualization software. However, there are apprehensions that Broadcom might be utilizing its dominant market position to extract more value from customers. Moreover, Broadcom's emphasis on immediate financial gains could result in a decrease in investments in research and development, potentially impeding VMware's ability to maintain a competitive edge and cater to the changing demands of its clientele. It is crucial to assess and potentially modify business strategies to capitalize on the strengths of both entities while addressing any discrepancies or obstacles in the integration process. This aspect is currently lacking and not apparent in the ongoing acquisition.

Overall, a thoughtful and deliberate approach to aligning business models and cultures will be essential for ensuring a successful transition and unlocking the full potential of the Broadcom-VMware combination. Since it is still not clearly evident even after many months of the merger. This is leading to distraught and high dissatisfaction by the VMware product users in the industry.

1.1.5 Impact on the Industry

The acquisition of VMware by Broadcom has implications for VMware's customers and partners. Here are some key considerations for these stakeholders.

1.1.5.1 Product Integration

Following the acquisition, both customers and partners have encountered changes in the integration and support of VMware's product portfolio. For instance, the VMware Aria Operations Cloud SaaS, vSphere+, and other similar products that were launched in 2023 to host applications in VMware Cloud have been discontinued in 2024 as a result of the merger. This has led to complications for numerous organizations, which are now required to relocate their deployments back to their original platforms. Many of the VMware products like these have lost their future post the Broadcom acquisition. It is crucial to remain updated on any modifications to VMware's product roadmaps and support procedures and to evaluate how to optimize VMware investments within the framework of the new combined entity.

1.1.5.2 Pricing and Licensing

Following the acquisition, Broadcom has implemented significant changes to VMware's pricing and licensing models. For instance, Broadcom has introduced a new minimum core purchase requirement of 16 cores per socket for servers. This means that even if a server has 2 sockets with 10 cores each, customers are now required to purchase licenses for a total of 32 cores under the new subscription-based model. The increase in the number of cores that customers must pay for has a substantial impact on the overall cost, as they are now obligated to pay for 32 cores instead of the previous 20 cores for the same server. This shift in pricing has come as a surprise to VMware consumers, as they are being charged for cores that they do not actually utilize. It is crucial for all parties involved to anticipate and potentially negotiate any alterations to pricing, licensing, or support agreements. Typically, enterprises plan their budgets before the start of the financial year, and the unexpected rise in costs introduced by Broadcom during the year is causing financial strain for many small to mid-sized organizations.

1.1.5.3 Technical Support

Subsequent to the merger, there have been alterations in the technical support protocols, as outlined in various online publications, due to the merging of the technical teams of the two companies. A number of clients have expressed dissatisfaction with the delayed response to their support inquiries following the merger. Consequently, it is advisable for customers and partners to remain flexible in accommodating any modifications in the support procedures and to effectively communicate their requirements to the relevant Broadcom technical teams.

1.1.5.4 Roadmap Alignment

VMware had a number of products that were of specific importance with many customers. For example, after the Broadcom acquisition, the VMware Horizon product was discontinued. Similarly, VMware ROBO (remote office/branch office), which was a very convenient and cheap virtualization offering by VMware for edge offices, is also now discontinued. In March 2024, Broadcom announced it would combine Symantec, which it purchased in 2019, with Carbon Black to create a new business unit called enterprise security group. Considering the changes Broadcom may further bring, users are required to adjust the product roadmap post-acquisition, and customers and partners should stay informed about any changes that may impact their businesses, such as changes in feature availability, product direction, or platform support. Customers and partners who were planning to adapt VMware technologies should now anticipate and prepare for potential changes in migration paths and transitional options brought about by the acquisition. They may need to reassess their existing environments and adjust their migration strategies or evaluate alternative options.

1.1.5.5 Communication

Due to the recent turbulence and layoffs at VMware, the dedicated account representatives who were previously in place have either been reassigned or are now shared among multiple organizations. This has had a significant impact on the transparent and open communication between VMware, Broadcom, and their respective stakeholders regarding any potential changes resulting from the acquisition. For instance, Broadcom offers a range of bundled products such as VMware Standard, VMware vSphere Foundation (VVF), and VMware Cloud Foundation (VCF). Among these offerings, VCF stands out as the most expensive bundle, encompassing a diverse array of products. However, it has been observed that Broadcom-VMware representatives are being pressured to exclusively promote the VCF offering. Numerous customers have raised concerns that Broadcom representatives are providing misleading information and failing to recommend suites that align with their specific needs (e.g., the VVF or Vsphere Standard suite). The rationale given for this is that the VVF bundle is intended only for small and retail businesses. Additionally, it has been suggested that the support provided with VVF suites is limited to Level 1 and Level 2 support. Nobody knows if these statements are true or to what extent. Therefore, considering this lack of clear communication, large enterprises are finding themselves compelled to opt for the costly VCF offering. As a result, it is imperative for customers and partners to maintain ongoing communication with VMware and Broadcom in order to stay informed, offer feedback on how to strengthen the partnership, and address any issues or concerns that may arise.

Overall, it is essential for customers and partners to prepare for potential changes and maintain close contact with Broadcom during the post-acquisition phase. By staying informed and anticipating potential changes, customers and partners can successfully navigate potential challenges and continue to receive value from their VMware investments through the new entity.

1.2   The Landscape of Previous Broadcom Mergers

Broadcom has transitioned from being primarily recognized for its networking chips to establishing itself as a major player in the rapidly evolving technology sector. Under the leadership of CEO Hock Tan, the company has implemented an assertive acquisition approach, resulting in its expansion across various industries, ranging from semiconductors to software. Through a series of strategic mergers and acquisitions, Broadcom Inc. has solidified its position as a prominent entity in the semiconductor and software market. In the last decade, the company has executed numerous notable acquisitions that have significantly broadened its product offerings and market presence. The history of Broadcom's acquisitions provides insight into the company's overarching strategy regarding the products and technologies it integrates. This section will delve into some of the key Broadcom acquisitions and their impact on the industry as a whole.

1.2.1 2015: Avago Technologies

Broadcom's initial significant purchase was Avago Technologies, a trailblazer in analog and mixed-signal semiconductors. On May 28, 2015, chip maker Avago Technologies Ltd. agreed to sell to Broadcom Corporation for $37 billion in cash and stock. This transaction reshaped Broadcom into a worldwide frontrunner in RF (radio frequency), optical, and connectivity solutions. The impact of this acquisition on the industry was profound, resulting in heightened competition, altered market dynamics, and a surge in innovation. The merged entity's robust financial standing and extensive product portfolio enabled it to actively pursue acquisitions and broaden its market presence, further solidifying its dominance in critical sectors. The merger also brought about a more consolidated semiconductor sector, with fewer players and mounting pressure on smaller firms to adapt or face acquisition. This acquisition acted as a catalyst for a wave of mergers and acquisitions in the technology sector, reshaping the competitive environment and laying the groundwork for future industry consolidation. The combined company emerged as a key player in the semiconductor field, offering a diverse array of products utilized in various devices such as smartphones, tablets, and computers. This consolidation of market influence empowered Broadcom to increase prices and diminish competition, ultimately resulting in higher costs for end consumers.

The acquisition had a profound effect on the cost of optical components, particularly on the price of optical transceivers. Following the acquisition, Broadcom emerged as the leading provider of optical transceivers, essential for data transmission through optical fiber cables. Subsequently, Broadcom implemented a substantial price hike of up to 50% on its transceivers. This surge in prices reverberated across the industry, compelling other suppliers to also raise their prices to stay viable in the market.

The Broadcom acquisition also led to a reduction in innovation in the optical component market. Before the acquisition, Avago Technologies was a major innovator in the development of new optical technologies. However, after the acquisition, Broadcom shifted its focus to cost-cutting and maximizing profits. This led to a decline in investment in research and development, which slowed the pace of innovation in the optical component market. The rise in prices coupled with a decline in innovation has adversely affected consumers. They are now facing increased costs for optical components without receiving the expected level of technological advancements that would have been possible if Avago Technologies had stayed independent. Consequently, end users are finding it challenging to stay abreast of the latest technological developments, thereby impeding the expansion of the optical component market.

1.2.2 2016: Brocade Communications Systems

The purchase of Brocade enhanced Broadcom's presence in the networking sector. Brocade was known for its expertise in data center switching, routing, and virtualization technologies. This $5.9 billion agreement solidified Brocade's leading position in Fiber Channel storage area networks (SANs) and data center networking, combining it with Broadcom's proficiency in semiconductor design and networking technologies. The acquisition resulted in significant market changes, as Broadcom integrated Brocade's SAN business into its existing portfolio and sold off the networking division to Extreme Networks. This action further strengthened Broadcom's dominance in the semiconductor market, providing a wide range of networking solutions, such as switches, routers, and Fiber Channel adapters. Meanwhile, the sale of the networking business to Extreme Networks enhanced its portfolio and enabled it to compete more effectively in the data center networking arena. The impact of the acquisition reverberated throughout the industry, prompting major players like Cisco to adjust their strategies and focus on emerging technologies like software-defined networking (SDN) and network functions virtualization (NFV). The acquisition also elicited mixed reactions, with some industry experts expressing concerns about Broadcom's potential to stifle competition and raise prices.

The merger of these two prominent companies in the networking sector has raised worries about diminished competition, potentially elevated prices, and restricted options for consumers. This is especially troubling given that Brocade was a major supplier of Fibre Channel networking solutions, a critical technology for storage area networks (SANs) utilized by numerous businesses. The assimilation procedure has encountered various difficulties, such as discontinuation of products, support challenges, and an overall feeling of uncertainty among customers who now depend on Broadcom for their networking requirements. Despite Broadcom's assurance that it will uphold support for Brocade's products, there are apprehensions that the integration process might result in decreased investment in these technologies, ultimately affecting the durability and dependability of existing infrastructure. The repercussions on end users, marked by reduced competition, potential price hikes, and ambiguity surrounding support and product plans, stand as a cautionary example for the future of industry consolidation.

1.2.3 2017: Qualcomm's Wireless Infrastructure Business

Broadcom's purchase of Qualcomm's wireless infrastructure division has bolstered its standing in the mobile network infrastructure sector. This transaction encompassed cutting-edge technologies for LTE, 5G, and Wi-Fi. The deal, valued at around $30 billion, has established a significant presence for Broadcom in the wireless infrastructure industry, offering a wide array of products and services. By acquiring Qualcomm's wireless infrastructure business, Broadcom has cemented its status as a top provider of chipsets, software, and services for wireless networks. Furthermore, this acquisition has expedited Broadcom's efforts to broaden its footprint in the wireless infrastructure market, which was at that time projected to experience substantial growth in the coming time. With this merger, the combined entity is now in a strong position to compete with key industry players such as Ericsson, Nokia, and Huawei.

By removing Qualcomm as a major competitor, Broadcom has achieved a dominant position, which could potentially allow it to dictate pricing and restrict options for service providers. This lack of competition has resulted in higher expenses for end users, who have experienced rising prices for mobile data, internet services, and other wireless products. Additionally, with Broadcom's increased control over a larger portion of the market, there is a possibility of diminished innovation as the company may prioritize its own financial interests over the advancement of state-of-the-art technologies. This could impede the progress of faster, more dependable, and more efficient wireless networks, which are essential for the development of various industries and the overall digital landscape. Ultimately, the acquisition has the potential to create a less dynamic and less consumer-friendly environment for wireless technologies, potentially leading to reduced options, increased expenses, and slower innovation.

1.2.4 2018: CA Technologies

The purchase of CA Technologies cemented Broadcom's position as a provider of enterprise software solutions. CA offered products for IT management, security, and DevOps. Broadcom, a semiconductor and infrastructure software company, gained access to CA's portfolio of IT management, security, and automation software, significantly enhancing its position in the enterprise software space. The deal had a profound impact on the industry, reshaping the competitive landscape and offering potential benefits and challenges for customers and stakeholders.

The acquisition of CA Technologies by Broadcom was primarily driven by financial interests rather than a genuine desire to create synergies. This has again led to a worrying trend of increased prices, reduced product quality, and diminished customer support. For example, after the acquisition, Broadcom significantly raised the cost of CA Technologies' products, forcing businesses to either pay more for the same services or switch to potentially less reliable alternatives. The quality of the products has also suffered, with features being removed or downgraded and updates being released with bugs and inconsistencies. As a result, businesses relying on CA Technologies' software have experienced increased downtime and decreased productivity. Additionally, the merger has resulted in a significant reduction in customer support, with long wait times, unresponsive staff, and limited knowledge about the acquired products. This has left customers feeling neglected and frustrated, especially when facing critical issues or needing urgent assistance. In essence, the Broadcom acquisition of CA Technologies, motivated by financial gain rather than a commitment to customer satisfaction, has created a negative ripple effect, ultimately impacting the end users who rely on these products for their business operations.

1.2.5 2019: Symantec's Enterprise Security Business

Broadcom's acquisition of Symantec's Enterprise Security business marked a significant expansion of its cybersecurity focus. The $10.7 billion deal granted Broadcom control over a substantial portion of Symantec's security solutions, including Endpoint Protection, Data Loss Prevention, and Web Security. With these acquisitions as well, initial concerns arose regarding potential price increases and reduced competition. However, the subsequent integration of the business has produced mixed outcomes for users. On one hand, the consolidation of security offerings has simplified security tool management for large organizations, potentially leading to cost savings and increased efficiency. Some users have also noted improved features and functionalities in the integrated products, reflecting Broadcom's investments in research and development. Conversely, the acquisition has faced criticism for its emphasis on cost-cutting measures, resulting in layoffs and possible service disruptions. Users have expressed worries about the impact on product support and innovation, as Broadcom's focus appears to be profit maximization rather than fostering a collaborative, customer-centric environment. The long-term effects on end users remain uncertain, despite some benefits from the integration. Concerns about reduced innovation and a profit-oriented approach raise doubts about the future of Symantec's enterprise security products under Broadcom's ownership. The impact on end users will largely hinge on Broadcom's dedication to upholding the quality and functionality of its inherited security offerings while balancing profitability and customer satisfaction.

From the past Broadcom acquisition, the following is the emerging inference for the end users and industry. This ambitious acquisition strategy is clearly driven by the following factors.

1.2.5.1 Consolidation Is the Trend

Broadcom has a history marked by a sequence of strategic acquisitions, demonstrating a strong dedication to consolidation in the technology sector. These mergers have played a significant role in consolidating the technology industry, positioning Broadcom as a major player. The company's goal is to establish a vertically integrated ecosystem, overseeing critical components and software solutions. This strategy has proven highly successful, propelling Broadcom from a specialized semiconductor firm to a giant encompassing various technologies such as networking, storage, software, and wireless communications. Despite the consolidation trend raising concerns about potential market dominance and its impact on innovation and competition, Broadcom's aggressive acquisition approach has undoubtedly transformed the tech industry, influenced the competitive landscape, and driven technological progress forward.

1.2.5.2 Improve Financial Performance

Broadcom has strategically utilized acquisitions to enhance its financial performance. The company's history is marked by a series of acquisitions aimed at expanding its product portfolio, increasing revenue streams, and optimizing its cost structure. By focusing on targeted acquisitions in the semiconductor industry, Broadcom has sought to drive growth, improve efficiency, and strengthen its competitive advantage. As a result of these acquisitions, Broadcom has experienced significant growth in revenue and profit margins. Acquisitions have enabled the company to streamline operations and achieve economies of scale. Broadcom's continued commitment to strategic acquisitions is driven by its financial objectives, which include increasing revenue, optimizing costs, and enhancing its competitive positioning in the semiconductor industry. Through these targeted acquisitions, Broadcom has established itself as a leading technology company and a key player in the global supply chain for semiconductors and connectivity solutions.

Broadcom's aggressive consolidation strategy has significantly transformed the technology landscape. The company's strategic acquisitions have led to a reduction in competition and the evolution of products and services tailored for niche markets where other providers have limited presence. This approach allows Broadcom to invest less in research and development while capitalizing on established niche markets. Consequently, Broadcom's position as a niche technology provider may lead to vendor lock-in, compelling customers to pay higher costs for the same level of services and support. Customers may feel obligated to pay due to the lack of alternative platforms for niche offerings, resulting in long-term dependencies on Broadcom's product portfolio. This, in turn, leads to financial gain for Broadcom but may drain the technological innovation in the industry. While the impact of this strategy is subject to debate, Broadcom's role in driving industry consolidation is undeniable. As the technology industry continues to evolve, Broadcom's influence will play a crucial role in shaping various sectors in the future.

1.2.6 Concerns About Broadcom's Reputation for Cost-Cutting, Impact on R&D, and Anti-competitive Practices

Broadcom Inc., a global semiconductor and infrastructure software company, has been scrutinized for its reputation for aggressive cost-cutting measures. This has raised concerns about the potential impact on its research and development (R&D) capabilities, which are crucial for maintaining technological leadership and innovation.

1.2.6.1 Reputation for Cost-Cutting

Broadcom is known for its track record of acquiring firms and executing substantial cost reduction initiatives. A case in point is the aftermath of the acquisition of Brocade Corporation in 2016, where the merged organization slashed numerous positions and shifted manufacturing operations to countries with lower labor costs. These measures have contributed to Broadcom's reputation as a "cost-killer."

Although the cost-cutting strategies implemented by Broadcom have positively impacted the company's financial performance, they have also sparked apprehension among employees and industry observers. Detractors contend that Broadcom's emphasis on cost-cutting has resulted in reduced innovation, diminished employee morale, and a deterioration in customer service. The company's history of layoffs and downsizing has also prompted inquiries into its dedication to retaining employees and fostering long-term growth. Nevertheless, Broadcom asserts that its cost-cutting measures are essential for optimizing efficiency, boosting profitability, and positioning the company for future success.

1.2.6.2 Concerns About R&D and Impact on Innovation

Broadcom's emphasis on cost-cutting has been criticized for potentially undermining its research and development efforts. R&D is a crucial but costly investment for technology firms, enabling them to innovate, improve existing products, and maintain a competitive edge. Overemphasis on cost reduction may result in sacrificing long-term innovation. Furthermore, the company's heavy reliance on acquisitions rather than internal R&D could lead to a lack of expertise and core competencies. A decrease in R&D expenditure could impede Broadcom's ability to develop state-of-the-art products and technologies, impacting various industries such as semiconductors, networking equipment, and software solutions. Without sustained R&D investment, Broadcom risks falling behind competitors who prioritize innovation, potentially eroding its market share and limiting long-term growth prospects.

Broadcom has shown commitment to investing in research and development, particularly in areas related to its acquired technologies. However, there are concerns regarding the potential long-term implications of its approach. Some critics suggest that Broadcom's emphasis on short-term financial gains and the integration of acquired technologies could hinder its ability to drive truly innovative and groundbreaking R&D initiatives. These critics highlight the company's track record of downsizing and reducing research budgets in acquired firms, which could result in a loss of talent and a possible decline in long-term innovation. On the other hand, proponents of Broadcom argue that its acquisition strategy enables rapid expansion of its product offerings and entry into new markets, thereby facilitating R&D through access to a broader array of technologies and expertise. Additionally, supporters contend that Broadcom's focus on commercialization and integration of acquired technologies accelerates product development and deployment, ultimately benefiting consumers and the industry.

The impact of Broadcom's acquisitions on its R&D reputation is a multifaceted issue. While the company's aggressive acquisition strategy has undoubtedly driven growth, the lasting effects on innovation and research are yet to be fully understood.

1.2.6.3 Reduced Competition

Broadcom's market dominance in specific sectors has the potential to hinder both innovation and price competition. The company's approach, commonly referred to as "buy and bury," entails the acquisition of competitors in critical markets, which could impede innovation and result in price hikes. For instance, the acquisition of VMware, a prominent virtualization software provider, has sparked concerns among antitrust authorities and industry analysts. While Broadcom asserts that its acquisitions enhance efficiency and foster innovation, critics highlight the consolidation of market share and the possible marginalization of smaller players. The company's reputation within the industry and among customers is often associated with its acquisition strategies, with some perceiving Broadcom as an aggressive consolidator, while others commend its emphasis on technology and integration. The ongoing discourse surrounding Broadcom's acquisitions and their impact on competition is expected to persist as the company continues to expand its presence in the technology sector.

1.2.6.4 Potential for Anti-competitive Practices

Broadcom's market dominance has raised concerns regarding its potential to harm competitors. The company's growth has been fueled by a series of aggressive acquisitions, drawing criticism for its alleged anti-competitive behavior. Critics have pointed to Broadcom's history of acquiring key players in various industries, such as Avago Technologies, which has allowed it to establish a strong presence in networking and storage chips. This consolidation has sparked fears of increased prices, limited consumer options, and reduced innovation, as Broadcom may exploit its market power. Moreover, accusations of predatory pricing and unfair competition tactics have further fueled apprehensions about Broadcom's conduct. The company's aggressive negotiation strategies, along with its pursuit of companies with complementary technologies, have led to claims of monopolistic practices and hindrance of alternative solutions. While Broadcom defends its acquisitions as a means to enhance its product offerings and foster innovation, its contentious approach and legal disputes with regulators and rivals have tarnished its reputation, resulting in allegations of anti-competitive conduct. The ongoing discourse on Broadcom's acquisition strategy underscores the intricate balance between business expansion, competition, and antitrust regulations, with potential implications for innovation and consumer welfare being central to the debate.

To conclude, irrespective of the concerns that loom around the Broadcom reputation around cost-cutting, potential effect on innovation, and anti-competitive practices, Broadcom is expected to continue targeting companies that complement its existing portfolio or provide access to new markets. The company's strong financial position and track record of successful integrations make it a formidable acquirer. As the semiconductor and software industries continue to evolve, Broadcom's acquisition spree is likely to shape its future trajectory. The company's goal is to become a leading provider of end-to-end solutions that connect devices, networks, and clouds.

1.3 Market Fallout: Dell Ends Agreement with VMware After Broadcom Acquisition and Emerging Rivals

1.3.1 Dell–VMware Relationship

Dell and VMware have enjoyed a long-standing and mutually beneficial relationship. VMware, a supplier of cloud computing and on-premises virtualization software, has been an essential component of Dell's enterprise solutions portfolio since Dell's acquisition of EMC, VMware's parent company, in 2016. Following the acquisition, Dell integrated EMC's enterprise storage, networking, and security divisions with its client solutions and peripherals business, establishing a comprehensive ecosystem of converged and hyper-converged solutions centered around products developed by Dell Technologies. By retaining VMware, Dell has continued to reap the benefits of the revenue generated by VMware's robust product offerings and its extensive customer base of more than 500,000.

1.3.2 Broadcom Enters the Game: End of the Dell–VMware Era

The acquisition of VMware by Broadcom has resulted in a change in the dynamics of the partnership between Dell and VMware. Dell Technologies has formally terminated its longstanding agreement with VMware, signifying a notable transformation in the technology industry subsequent to Broadcom's acquisition of VMware. This development, disclosed in January 2024, involved Dell Technologies submitting a notice to the Securities and Exchange Commission (SEC) (`https://investors.delltechnologies.com/node/15351/html`).

This is to say the company is officially ending its distribution agreement with VMware. In the filing, Dell announced that it had delivered to Broadcom the written notice of termination of the commercial framework agreement that was initiated in November 2021. Dell noted that according to the contract signed, either party could end the agreement within 60 days of new ownership – or "Change of Control" – which is what happened in the VMware acquisition. This comes after months of speculation and negotiation between the two companies, fueled by Broadcom's aggressive acquisition strategy and Dell's desire to maintain its independence. The cancelled Dell–VMware agreement set a framework for several commercial activities and collaborations between the two parties. The contract also called for continued partnership on several VMware products, including the VxRail hyper-converged infrastructure (HCI) platform, VMware Cloud on Dell EMC, VeloCloud SD-WAN, and more. While Dell and VMware have enjoyed a mutually beneficial partnership for decades, the acquisition presented a complex situation. Dell, initially a committed supporter of the acquisition, expressed concerns about potential conflicts of interest and the impact on customer choices. The agreement termination signifies Dell's commitment to its own technology platform and its desire to establish a more independent path forward. This move allows Dell to pursue alternative virtualization and cloud solutions, including its own offerings and those of other partners. Furthermore, it allows Dell to foster stronger relationships with other key players in the technology ecosystem, fostering competition and innovation. The termination of the agreement signals a significant transition in the relationship between Dell and VMware, potentially opening doors for new collaborations and partnerships as both companies navigate the evolving landscape of technology and cloud computing.

Following Broadcom's $61 billion acquisition of VMware in May 2023, Dell re-evaluated its strategic partnership with VMware, particularly in light of Broadcom's reputation for cost-cutting measures and its emphasis on hardware and software integration. Dell had previously maintained a strong alliance with VMware, utilizing VMware's virtualization solutions to complement its own hardware offerings and deliver comprehensive data center solutions to customers. However, Dell's concerns revolved around the potential disruption to its current business model, the risk of increased hardware costs as a result of Broadcom's focus on hardware sales, and the potential impact on VMware's open and multi-cloud approach under Broadcom's ownership. Additionally, Dell's decision reflected its broader strategy of diversifying its product portfolio and reducing dependence on third-party software vendors in order to maintain control over its hardware and software stack. This move is in line with Dell's vision of evolving into a more vertically integrated technology solutions provider, offering customers a comprehensive range of products and services without the limitations of external partnerships.

Dell's decision to separate from VMware is aimed at streamlining its business structure and operations, with a focus on its core areas of expertise. The termination of the agreement will enable Dell to allocate resources and capital more effectively toward its infrastructure and peripherals businesses. Furthermore, the split will provide an opportunity for Dell to reassess and enhance its go-to-market strategies for its server, storage, and networking solutions. This may result in more precise and efficient sales approaches, ultimately driving growth in these segments and delivering value to Dell's investors.

1.3.3 Dell's VXRAIL Discontinuation: A Post-merger Shift or Strategic Realignment?

Before jumping into the topic, let us quickly highlight the terminology hyper-converged infrastructure (HCI) platforms and provide a brief about VXRail platform.

According to the wiki definition:

> *Hyper-converged infrastructure (HCI) is a* software-defined IT infrastructure *that virtualizes all of the elements of conventional* "hardware-*defined*" *systems. HCI includes, at a minimum,* virtualized computing *(a* hypervisor*), software-defined storage, and virtualized networking* (software-defined networking*). HCI typically runs on* commercial off-the-shelf *(COTS) servers.*

Figure 1-1 provides some quick graphical images that distinguish the three different popular technology stacks available.

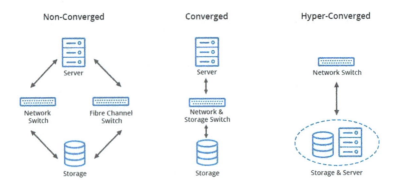

Figure 1-1. *Difference between non-converged, converged, and hyper-converged network storage*
(Source: https://en.wikipedia.org/wiki/Hyper-converged_
infrastructure#/media/File:Hyperconvergence.jpg*)*

In a non-converged environment, administrators historically managed the three core infrastructure components separately, procuring equipment from various technology solution providers such as storage from EMC, NetApp, Nutanix, etc.; networking from Cisco, Juniper, etc.; and servers from Dell, HPE, Cisco, Lenovo, etc. This operational approach persisted until the rise of converged infrastructure, where multiple vendors collaborated to deliver a unified solution that conceals the underlying complexity of compatibility requirements between different technology stack components when integrated together. With converged solutions, the integration and compatibility responsibilities are transferred to the technology vendors, who conduct compatibility testing in their labs and furnish end users with a supported compatibility matrix of their product releases for deployment in their own infrastructure stack. An example of a prominent converged solution in the market is Dell VxBlock, which offers end users the flexibility to select their preferred technology partners for storage, compute, and network, providing them with a single, ready-to-deploy rack that can scale according to customer requirements.

Considering the advantage with hyper-converged solution, where the three core infrastructure aspects (storage, compute, and network) are consolidated into one single solution and there is a single pane of glass for the management of the necessary core components. The popularity of hyper-converged solutions has grown significantly. Following are key reasons for this:

1. The integration provided by an HCI platform results in decreased space, cooling needs, and lower TCO (total cost of ownership). This is due to the fact that HCI solutions are usually deployed on a unified hardware that can deliver software-defined storage, networking, and computing capabilities.

Consequently, there is no necessity for distinct rack space for various essential elements. Additionally, the complexities associated with intricate cabling and rack space allocation are eliminated with HCI product technology.

2. HCI is typically offered by a single technology vendor; thus, in case of support issues, customers only have to deal with one vendor instead of multiple vendors in "non-converged" and "converged" solution offerings.

3. A single management plane offered with HCI is another great addition that helps infrastructure administrators tremendously. This is due to their ability to effectively manage infrastructure provisioning, upgrades, and maintenance using a single interface that allows for seamless navigation between different core technologies.

Dell VxRail is also an HCI system that combines compute, storage, networking, and virtualization resources into a single device. This platform is jointly engineered by Dell EMC and VMware organizations. It is a fully integrated, preconfigured, and tested HCI system optimized for VMware vSAN software-defined storage and the VMware vSphere ESXi hypervisor.

Dell's recent decision to halt the production of the VxRail hyper-converged infrastructure (HCI) platform has caused a stir within the IT industry. This move follows closely after Broadcom's acquisition of VMware, marking a significant shift in the virtualization and HCI solutions market. Although Dell has not explicitly connected these two occurrences, the timing and the changing market conditions indicate a multifaceted relationship between various factors.

1.3.3.1 The Broadcom-VMware Deal: A Turning Point

The acquisition of VMware by Broadcom has sparked concerns regarding potential shifts in VMware's product roadmap and pricing strategies. Speculation among industry analysts suggests that Broadcom may prioritize its own software offerings over those of VMware, potentially resulting in diminished innovation and higher costs for VMware users. This uncertainty surrounding VMware's future has undoubtedly played a role in influencing Dell's decision-making process. Given that VxRail, a critical component of Dell's HCI strategy, heavily relies on VMware's vSphere software, the possibility of alterations in VMware's future trajectory could have posed a threat to Dell's longstanding commitment to the platform. Consequently, end users and organizations heavily invested in Dell–VMware products may be impacted by these developments. The organization's future trajectory, previously centered around the VxRail hyper-converged market, will likely undergo significant evolution as it explores alternative solutions in the market.

1.3.3.2 Dell's Focus Shift: A Strategic Realignment?

While the VxRail discontinuation might appear sudden, it also aligns with Dell's broader strategic focus on edge computing, cloud-native services, and software-defined infrastructure. Dell is working on developing solutions such as Dell PowerEdge servers and the Apex platform to address these evolving trends. Additionally, the discontinuation may indicate Dell's intention to streamline its product lineup and prioritize solutions that provide a stronger competitive edge. As alternative HCI platforms emerge and cloud-based solutions gain traction, Dell could be strategically realigning its focus toward offerings that are more in tune with future market needs.

1.3.3.3 Implications for HCI (Hyper-converged Infrastructure) Market

Dell VxRail is widely recognized as a leading provider of HCI solutions, integrating VMware's robust virtualization technology with top-tier storage and compute hardware from Dell and EMC. The decision to discontinue VxRail by Dell carries substantial consequences for the HCI sector. Although Dell's dedication to HCI persists, this development hints at a possible transformation in the competitive environment. Competitors such as Microsoft, Nutanix, HPE, and Cisco could capitalize on this situation to enhance their market presence.

The termination of VxRail by Dell serves as a reminder of the significance of adaptability and flexibility in the rapidly changing IT industry. As technology continues to advance at an unprecedented rate, it is imperative for vendors to continuously evaluate their offerings and modify their strategies in order to stay competitive. Dell's choice to discontinue VxRail reflects the shifting landscape, indicating a shift toward a more dynamic and flexible approach to infrastructure management, where adaptability and agility are crucial. In the foreseeable future, we can anticipate further consolidation and innovation in the HCI market as vendors strive to meet the evolving demands of businesses in a world increasingly influenced by cloud and edge computing. Dell's decision may mark the beginning of a new era in HCI, one in which flexibility, scalability, and cloud-native solutions take precedence.

1.3.3.4 Emerging Rivals in the Wake of VxRail Discontinuation

Dell EMC's decision to discontinue its VXRAIL hyper-converged infrastructure (HCI) solution has created an opening in the HCI market, allowing new competitors to enter the scene. These emerging rivals are eager to take advantage of the opportunity to offer alternative HCI solutions that cater to the changing needs of customers. Notable contenders in this space include Nutanix, Microsoft, and Cisco. Nutanix, with its Nutanix Acropolis Hypervisor (AHV) and established HCI offering, is well-positioned to capture a larger market share. Microsoft, on the other hand, has introduced its Azure Stack HCI platform, which is based on the Hyper-V virtualization engine. Cisco, leveraging its networking expertise, is also making a foray into the HCI market with its Hyperflex system, aiming to provide customers with a comprehensive end-to-end solution. With these competitors ramping up their efforts, the HCI market is expected to undergo a period of rapid innovation and consolidation. Customers in search of HCI solutions will now have a wider array of options, as each vendor strives to deliver the best combination of performance, reliability, and cost-effectiveness. The discontinuation of VxRail has set the stage for a dynamic and competitive HCI market, where new rivals will challenge established players and drive the evolution of this crucial technology.

The competitors in the market exhibit a variety of strengths and weaknesses as they vie for market dominance. Nutanix, a well-established player, showcases a robust product lineup with its hyper-converged infrastructure solutions, providing adaptable scalability and a wide range of features. Despite Nutanix's extensive customer base and strong partner network, its pricing may pose a challenge for certain customers. Microsoft's Azure Stack HCI platform excels in offering Azure Cloud as a management platform for their on-prem HCI infrastructure, setting them apart for customers seeking cloud-like management capabilities and cloud-native automations for on-premises deployments. Cisco HyperFlex stands out with its impressive networking capabilities and seamless integration with Cisco's broader product range, making it an appealing choice for organizations already utilizing Cisco infrastructure. However, HyperFlex may not offer the same level of hardware flexibility as some of its competitors. Dell Technologies leverages its diverse hardware portfolio and solid customer relationships to deliver a competitive HCI solution utilizing the PowerFlex hardware platform. Lenovo ThinkAgile provides a cost-effective and streamlined HCI solution, particularly appealing to smaller organizations. Nevertheless, ThinkAgile may lack the extensive features found in larger competitors. Despite its strong presence in the Asia-Pacific region, Huawei FusionCube encounters difficulties in expanding its global market share due to political and geopolitical factors. HPE SimpliVity, while offering seamless integration with HPE's broader product range, faces challenges with declining market share. Each emerging competitor brings a distinct value proposition to the HCI market, showcasing their individual strengths and weaknesses. Their success in capturing market share will hinge on their ability to cater to the specific requirements of modern enterprises, striking a balance between features, performance, pricing, and security in a fiercely competitive environment.

1.3.3.5 The Future of Hyper-converged Infrastructure (HCI) Market

Hyper-converged infrastructure (HCI) has garnered substantial attention in the IT sector due to its myriad benefits compared to traditional infrastructures. Pertinent data, statistics, and research outcomes offer compelling evidence in favor of HCI adoption. As per research reports, the HCI market is forecasted to achieve a value of $10.2 billion by 2024, indicating a compound annual growth rate (CAGR) of 21.8% from 2018 to 2024. This growth is primarily fueled by the escalating demand for agile, scalable, and cost-efficient IT solutions. Research conducted by the IDC demonstrates that HCI can reduce IT complexity by 60%, resulting in significant operational cost savings. Through the consolidation of compute, storage, and networking into a unified, software-defined platform, HCI eliminates the necessity for multiple hardware components and intricate management tasks. Moreover, findings from a study carried out by Gartner reveal that HCI enhances application performance by 30%. The amalgamation of storage and compute resources within a single node diminishes latency and ensures consistent performance for resource-intensive applications. HCI also bolsters data availability and disaster recovery capabilities. A report from Forrester Research underscores that HCI has the potential to decrease data loss by 90%, owing to its inherent data protection functionalities like replication, snapshots, and automatic failover. Additionally, research by ESG Research indicates that HCI can reduce the total cost of ownership (TCO) by 30–40%. The eradication of hardware silos, simplified management processes, and decreased power consumption collectively contribute to substantial cost savings.

The data and investigations presented establish a robust groundwork for examining hyper-converged infrastructure (HCI). They illustrate the increasing acceptance of HCI in the market, the advantages in operations, improvements in performance, capabilities in data protection, and the potential for cost savings. Consequently, HCI is positioned to persist as a revolutionary technology for enterprises aiming to update their IT infrastructures and propel digital advancement. By amalgamating computing, storage, networking, and virtualization resources into a unified system, HCI offers enhanced simplicity, scalability, and cost efficiency. Consequently, HCI is gaining momentum across various sectors, especially in small and medium-sized enterprises (SMBs) and edge computing settings.

Currently, a prominent trend is the growing utilization of cloud-based HCI services. These services, which are based on cloud technology, offer benefits such as scalability, flexibility, and reduced capital expenditure. Businesses are increasingly opting for cloud-based HCI services to meet their IT infrastructure requirements, leading to market expansion. Moreover, there is a rising interest in composable infrastructure, allowing enterprises to allocate and adjust resources dynamically for improved agility and efficiency. Additionally, the emergence of software-defined networking (SDN) and software-defined storage (SDS) is impacting the HCI market by providing enhanced flexibility and programmability to hyper-converged infrastructure solutions, enabling customization to meet specific needs. The integration of software-defined networking and storage in a unified platform offers increased flexibility and scalability. The market is shifting

toward cloud-native HCI solutions, facilitating seamless integration with public cloud services and boosting hybrid cloud adoption. Furthermore, the growing adoption of containerization and microservices architectures is fueling the demand for HCI solutions tailored to these workloads.
The future of the HCI market is expected to involve the convergence of technologies like artificial intelligence (AI), machine learning (ML), and edge computing, driving innovation and improving business agility. These technologies are paving the way for the development of advanced HCI solutions that are more responsive to business requirements.

Overall, the HCI environment is currently experiencing a significant transformation due to the rising popularity of HCI, cloud-based CI services, composable infrastructure, and software-defined technologies. These changes and developments are anticipated to persist in altering the industry landscape, creating fresh prospects for suppliers and empowering businesses to enhance their IT infrastructure for improved flexibility, productivity, and cost-efficiency.

1.4   The Future Is Uncertain: What Now for VMware After Broadcom's Takeover?

The acquisition of VMware by Broadcom marks a notable shift for the organizations involved, with the full implications yet to be realized. The future trajectory of their collaboration remains ambiguous as they venture into uncharted territories to foster development and creativity. The effects of this merger on market competition and the progression of corporate technology in the foreseeable future are eagerly anticipated.

There are numerous questions in the minds of VMware proponents about the future and how this Broadcom merger is going to evolve. Let us try to anticipate future directions and what we should do to better align ourselves to better brace ourselves to meeting our ever-evolving business needs.

1.4.1 A Shift in Focus from Broadcom?

A major worry revolves around the possibility of Broadcom shifting its focus away from VMware's innovation trajectory. Broadcom, a renowned provider of networking and infrastructure solutions, might emphasize cost reduction and revenue generation at the expense of innovation and progress. This could have implications for the future direction of VMware's key products, such as vSphere and vSAN, and could potentially impact the frequency of new feature releases and bug fixes. While the future remains uncertain, it is important to be ready to adapt to any changes that may arise. It is crucial to recognize industry developments and realign our business strategies accordingly.

1.4.2 The Cloud Uncertainties

VMware has strategically broadened its range of offerings to compete with AWS, Azure, and GCP in response to the increasing prevalence of cloud computing. The acquisition by Broadcom raises questions about the impact on this strategy. It remains to be seen whether VMware will maintain its significant investment in cloud products or if the focus will shift toward on-premises solutions, potentially leaving cloud customers in a state of uncertainty. This merger introduces another dimension of uncertainty, but as the industry continues to evolve, it is essential to carefully consider all available options that align with our business needs.

1.4.3 The User Experience

Numerous individuals express apprehension regarding the possible effects on support and pricing. Will Broadcom give precedence to enhancing customer experience or emphasize cost-effectiveness, potentially resulting in alterations to support services or higher licensing fees? This trend is already manifesting itself, with customers now facing elevated prices for products they have been using for years. Looking ahead, the situation may deteriorate further if Broadcom continues to prioritize revenue generation over technological advancements. As consumers, it is advisable to prepare for any unforeseen circumstances by expanding our partnerships with various vendors to avoid being tied to a single supplier.

1.4.4 The Bigger Picture

The merger has the potential to significantly impact the virtualization market as a whole. Could rival companies perceive this as a chance to enhance their positions and occupy the space that VMware might leave behind due to decreased innovation? Might we witness market consolidation as smaller firms find it challenging to stay competitive? As end users, it is imperative for us to carefully assess various vendors available in the market to identify those that align with the business needs.

1.4.5 A Call for Transparency

Broadcom must engage in open and transparent communication to address concerns and establish trust regarding its future intentions for VMware. It is imperative to foster an open dialogue with users, developers, and the broader community in order to facilitate a seamless transition and mitigate any possible adverse consequences.

1.4.6 The Future Is Uncertain, but Not Hopeless

While the future of VMware under Broadcom's ownership is masked in uncertainty, it's not all doom and gloom. There could be chances that the deal could potentially bring financial stability and resources that could help VMware innovate and expand its offerings in new and exciting ways. However, it remains crucial for Broadcom to prioritize customer needs, innovation, and transparency to ensure a positive outcome for VMware's future.

Broadcom may be ignoring the market and end users' needs, but there are several emerging vendors that offer a more promising future in the industry where VMware operates. It is crucial for us to stay informed about the changing landscape and remain open to exploring alternative solutions.

1.4.7 What Does This Mean for You?

It's crucial for your business to stay vigilant and keep a close eye on this developing situation. Even though the initial effects might not be significant, the potential long-term outcomes could have an impact on your IT infrastructure decisions. Take the time to assess your current virtualization requirements and look into different options offered by various vendors.

1.5 Summary

The acquisition of VMware by Broadcom has caused significant disruption in the IT industry, leading to a great deal of uncertainty regarding the future of VMware products and innovations. This chapter delves into the various implications of this acquisition, exploring different perspectives and shedding light on past acquisitions by Broadcom. The discussion also examines the current Dell–VMware relationship and the impact on the Dell VxRail product. Furthermore, in the end the chapter introduced emerging HCI technologies that end customers can evaluate and utilize in response to the turmoil surrounding VMware.

CHAPTER 2

Unveiling the Broadcom-VMware changes

The primary goals of this chapter are to shed further information on and delve further into the Broadcom VMware transformations. This chapter's opening portion will cover VMware's previous licensing schemes and contrast them with the most recent modifications made to the company's product offerings by Broadcom, specifically the addition of subscription-based solutions and core-based computations. We'll also go into detail about the various bundles that are on sale and compare them all, along with the corresponding "add-ons" that go along with each bundle. We will provide some of the cost figures for the various bundles that are offered later in the chapter, based on the various customer blogs and experiences. Lastly, we will bring to light some of the negotiation tactics that allow customers to be equipped and ready when they want to go into the negotiations with VMware and can take informed choices.

© Sumit Bhatia and Chetan Gabhane 2025
S. Bhatia and C. Gabhane, *Navigating VMware Turmoil in the Broadcom Era*,
https://doi.org/10.1007/979-8-8688-1264-4_2

2.1 Deciphering New Licensing Model

In the preceding chapter, we provided a concise overview of the alterations in licensing models and bundles available from Broadcom after acquiring VMware. Let's now explore this topic further by examining the licensing model in place before and after the Broadcom acquisition. Following that, we will unravel the new licensing models and product bundles offered by VMware under Broadcom's management.

2.1.1 Background: Understanding VMware's Old Licensing Models

Before diving into the new licensing models, it's essential to understand the old framework. This would help to better understand the impact of the overall change that Broadcom has brought to the licensing. VMware had previously provided various licensing models tailored to meet customer needs. These models encompassed per-CPU, per-virtual machine, per-physical machine, and vRAM entitlements. Organizations were presented with a range of options to select from based on their infrastructure needs and budget constraints.

1. **Per-CPU licensing**: In this model, organizations used to purchase licenses based on the number of physical CPU sockets present in the servers. Each CPU socket requires a minimum of four licenses, regardless of the number of cores.

2. **Per-virtual machine licensing**: Here, organizations purchased licenses based on the number of virtual machines deployed in their infrastructure. Each virtual machine requires a license, regardless of the underlying hardware resources.

3. **Per-physical machine licensing:** VMware provided licensing options based on the physical host, with vSphere Enterprise Plus being the most popular choice. In this case, licensing is based on the number of processors and cores.

4. **vRAM entitlements**: VMware introduced vRAM entitlements with vSphere 5. Organizations purchased vRAM entitlements based on the vRAM capacity allocated to their virtual machines. However, vRAM entitlements were discontinued in vSphere 6.

The different licensing options were one of the major attractions of using VMware in the on-premises setup. The kind of flexibility that was offered was a key differentiator and possibly one of the significant reasons for VMware's popularity. From the past decade, these innovative licensing options have attracted a lot of customer base and led to the popularity of VMware in the industry.

With the old licensing model in mind, let us further dive into different bundles offered by Broadcom.

2.1.2 Broadcom VMware License Changes

Broadcom's decision to implement significant changes to VMware's licensing model has sparked considerable discussion within the IT community and among existing VMware customers. The shift comes after Broadcom's acquisition of VMware, which raised concerns about the future of VMware's flexible and user-friendly licensing structures that many organizations have come to rely on for efficient resource management and cost predictability. Let us understand these changes in detail.

2.1.2.1 Packaging and Bundle Changes of VMware Products

On December 11, 2023, Broadcom announced the discontinuation of sales and support/maintenance renewals for several VMware products, effective immediately. Additionally, Broadcom has made significant changes for VMware customers by ceasing the sale of individual VMware product licenses. Previously, customers could purchase individual VMware products such as Aria Operations, vSphere, NSX, etc., independently. However, following the acquisition, Broadcom is primarily promoting two major product bundles moving forward. These bundles include VMware Cloud Foundation (VCF) and VMware vSphere Foundation (VVF). There are other bundles as well, but they are not heavily promoted by Broadcom. We will discuss all the different bundles in detail later in this chapter. Also, what sets these bundles apart is that they combine some of the previously available independent VMware products into one package. For example, customers using NSX and vSphere would now need to purchase the expensive VCF bundle, as the VVF or VVS (VMware vSphere Standard) bundle does not include NSX as one of its components. Again, we will delve into the details of these bundles in the subsequent sections of this chapter; here we are just highlighting the license changes. The affected VMware products include those that customers are currently using, and they will need to convert their product licenses into VCF or VVF bundles and purchase subscription-based licenses available in 1-year, 3-year, and 5-year terms. On a high level, here is the list of VMware individual products available in the VMware Cloud Foundation (VCF) bundle.

- VMware vSphere
- VMware vSAN
- VMware NSX
- VMware HCX
- VMware Site Recovery Manager

- VMware vCloud Suite

- VMware Aria Suite

- VMware Aria Universal

- VMware Aria Automation

- VMware Aria Operations

- VMware Aria Operations for Logs

- VMware Aria Operations for Networks

Broadcom's claim for combining these products into one bundle is that this is

A dramatic simplification of our product portfolio that allows customers of all sizes to gain more value for their investments in VMware solutions. The portfolio simplification across all VMware by Broadcom divisions stems from customer and partner feedback over the years telling us our offers and go-to-market are too complex.

As a result, existing customers are required to convert their existing VMware purchases to the new subscription-based bundle product (such as VCF or VVF) that are only currently available for their purchase. Also, if the VCF and VVF bundles do not include the products customers use (e.g., Site Recovery Manager, NSX Firewall, etc.), then Broadcom is providing options to do an "add-on" with additional costs on top of their VCF or VVF bundle purchase.

As part of portfolio simplification, Table 2-1 is the comprehensive list of products no longer available as standalone offerings. The below table maps whether the product is mapped to any replacement product or add-on. If it is mentioned as "N" under replacement product, then that means that individual product is discontinued, and no alternative is available. In that case, customers need to move to a solution or offering that is similar or a higher-tier solution providing similar or advanced capabilities than those currently offered.

Table 2-1. *Mapping table for products no longer available as standalones and also informs if these products have replacement products available.*

Source: https://blogs.vmware.com/cloud-foundation/ 2024/01/22/vmware-end-of-availability-of-perpetual- licensing-and-saas-services/

Products no longer available as standalones (all editions and pricing metrics)	Replacement product included in VCF/VVF/ Add-On? (Y/N)	Which product or add-on?
VMware vSphere Enterprise Plus	Y	VCF, VVF
VMware vSphere+	N	
VMware vSphere Enterprise	N	
VMware vSphere Standard (excluding subscription)	Y	Replaced with new vSphere Standard
VMware vSphere ROBO	N	
VMware vSphere Scale Out	N	
VMware vSphere Desktop	N	
VMware vSphere Acceleration Kits	N	
VMware vSphere Essentials Kit	N	
VMware vSphere Essentials Plus (excluding new subscription offering)	Y	Replaced with vSphere Essentials Plus Kit
VMware vSphere Starter/ Foundation	N	
VMware vSphere with Operations Management	N	

(continued)

Table 2-1. (*continued*)

Products no longer available as standalones (all editions and pricing metrics)	Replacement product included in VCF/VVF/ Add-On? (Y/N)	Which product or add-on?
VMware vSphere Basic	N	
VMware vSphere Advanced	N	
VMware vSphere Storage Appliance	N	
VMware vSphere Hypervisor (free edition)	N	
VMware Cloud Foundation/ VCF+/VCF-S (excluding new VCF subscription offering)	N	Replaced with VMware Cloud Foundation (new)
VMware Cloud Foundation for VDI	N	
VMware Cloud Foundation for ROBO	N	
VMware SDDC Manager	Y	VCF
VMware vCenter Standard	Y	VCF, VVF, and vSphere STD
VMware vCenter Foundation	N	
VMware vSAN	Y	VCF, VVF, vSAN add-on
VMware vSAN ROBO	N	
VMware vSAN Desktop	N	
VMware vSAN+	N	
VMware HCI Kit	N	

(*continued*)

Table 2-1. (*continued*)

Products no longer available as standalones (all editions and pricing metrics)	Replacement product included in VCF/VVF/ Add-On? (Y/N)	Which product or add-on?
VMware Site Recovery Manager	Y	SRM add-on service
VMware Cloud Editions/Cloud Packs	N	Replaced with VCF, VVF
VMware vCloud Suite	N	Replaced with VCF, VVF
VMware Aria Suite (formerly vRealize Suite)	Y	VCF, VVF
VMware Aria Universal Suite (formerly vRealize Cloud Universal)	N	
VMware Aria Suite Term	Y	VCF, VVF
VMware Aria Operations for Networks (formerly vRealize Network Insight)	Y	VCF
VMWare Aria Operations for Networks Universal (formerly vRealize Network Insight Universal)	N	
VMware vRealize Network Insight ROBO	N	
VMWare Aria Operations for Logs (formerly vRealize Log Insight)	Y	VVF, VCF

(*continued*)

Table 2-1. (*continued*)

Products no longer available as standalones (all editions and pricing metrics)	Replacement product included in VCF/VVF/ Add-On? (Y/N)	Which product or add-on?
VMware vRealize Operations 8 Application Monitoring Add-On	N	
VMware Aria Operations	Y	VVF, VCF
VMware Aria Automation	Y	VCF
VMware Aria Suite Cloud for US Public Sector	N	
VMware Aria Automation for Secure Hosts add-on (formerly SaltStack SecOps)	Y	Tanzu Guardrails Add-On
VMware vRealize Automation SaltStack SecOps add-on	Y	Tanzu Guardrails Add-On
VMware Aria Operations for Integrations (formerly vRealize True Visibility Suite)	Y	VCF, VVF
VMware Cloud Director	Y	VCF
VMware Cloud Director Service	N	
VMware NSX	Y	VCF and VMware Firewall
VMware NSX for Desktop	Y	VCF and VMware Firewall
VMware NSX ROBO	Y	VCF and VMware Firewall

(*continued*)

Table 2-1. (*continued*)

Products no longer available as standalones (all editions and pricing metrics)	Replacement product included in VCF/VVF/ Add-On? (Y/N)	Which product or add-on?
VMware NSX Distributed Firewall	Y	VCF and VMware Firewall
VMware NSX Gateway Firewall	Y	VCF and VMware Firewall
VMware NSX Threat Prevention to Distributed Firewall	Y	VCF and VMware Firewall
VMware NSX Threat Prevention to Gateway Firewall	Y	VCF and VMware Firewall
VMware NSX Advanced Threat Prevention to Distributed Firewall	Y	VCF and VMware Firewall
VMware NSX Advanced Threat Prevention to Gateway Firewall	Y	VCF and VMware Firewall
VMware Advanced Load Balancer	Y	VMware Avi Load Balancer Add-On (also standalone)
VMware Container Networking Enterprise with Antrea	Y	VCF and VMware Firewall
VMware HCX	Y	VCF
VMware HCX+	N	
Tanzu Kubernetes Grid (vSphere with Tanzu)	Y	VCF

Another important thing to note is that Broadcom discontinued the Aria SaaS services, which they started in the year 2022. Here is the statement from Broadcom in their blog.

> *VMware is also announcing EoA of the Aria SaaS services but will continue to support those customers currently using Aria SaaS services until the end of their subscription term. At the end of the subscription term, customers will need to transition to either VCF or VVF. Also note that Aria SaaS Services are in maintenance mode, meaning no new features will be published, although security and critical updates will be provided during the active subscription period.*

Customers who have been utilizing the Aria suite of products under an Aria SaaS subscription will now need to transition their deployments back to on-premises self-managed VM appliances within their own infrastructure. You can find more information regarding the changes to the Aria SaaS product portfolio discontinuation in the following blog: *https://blogs.vmware.com/management/2024/01/dramatic-simplification-of-vmware-aria-as-part-of-vmware-cloud-foundation.html*

2.1.2.2 Licensing Model Changes

After comprehending the modifications made to VMware products and bundles, it is imperative to shift our focus toward the alterations in the licensing model following the acquisition by Broadcom. The changes in the licensing model should be examined from two overarching perspectives. These aspects will be discussed in detail below.

a. End of Perpetual Licenses

A significant modification in the licensing terms was revealed on December 11, 2023, which includes the termination of the sale of VMware perpetual licenses. Companies holding perpetual licenses prior to this date can continue to use the versions they have acquired but will no longer have access to support or updates once their contract expires. This presents a significant risk in terms of security and long-term compatibility. Broadcom declared that they will exclusively offer subscription-based licenses with contract durations of 1 year, 3 years, and 5 years. Additionally, starting from that date, Broadcom stated that they will cease the sale of support and services (SnS) for perpetual licenses. The following excerpt is from Broadcom's blog addressing whether customers can renew their SnS contracts post Broadcom's acquisition.

> *Customers cannot renew their SnS contracts for perpetual licensed products. Broadcom will work with customers to help them "trade in" their perpetual products in exchange for the new subscription products, with upgrade pricing incentives.*

Broadcom also confirmed that it is not an obligation to "trade in" their existing perpetual licenses. It is only to get better pricing incentives when they are moved to the VCF or VVF bundle. Broadcom also confirmed that they would be honoring the existing SnS contracts for the durations they are applicable but no longer going to issue new SnS for the existing perpetual licenses. Here is the news blog that is published on the Broadcom website: *https://news.broadcom.com/news/vmware-by-broadcom-business-transformation*

b. Transition from CPU to Core-Based Pricing

The next significant aspect regarding the license modifications is the discontinuation by Broadcom of the CPU-based pricing model. Instead, they have implemented a core-based pricing structure. Furthermore, they have established a minimum requirement for the number of cores that a customer must purchase per physical CPU within the server. The minimum number of cores that must be acquired is 16 cores per socket. The fundamental calculation details are outlined below.

Foundation Subscription Licensing = The VCF and vSphere Foundation subscription capacity that requires licensing is the greater of either:

- Number of core licenses required per CPU × Number of CPUs per ESXi host × Number of ESXi hosts

- 16 cores × Total number of CPUs in each ESXi host

To further illustrate, in the case where a server is equipped with two physical CPUs, each containing 10 cores, the total number of physical cores available in that server is 20. Consequently, instead of acquiring subscription licenses for 20 cores, clients are now required to procure a minimum of 36 cores to match their current licensing obligation. This is because the minimum number of cores that must be purchased per physical CPU is 16. It is important to emphasize that the 16 cores per socket represent the minimum requirement for purchase. However, if the number of cores per socket exceeds 16, the exact core count must be considered for licensing purposes. For instance, if a server has 20 cores per socket and there are two physical CPUs in total, then a subscription for 40 cores would be necessary for that server.

There have been adjustments made to the vSAN license calculations. When a customer purchases the VMware Cloud Foundation (VCF) bundle, they receive a complimentary vSAN entitlement of 1 TiB per core of the license they acquire for the cluster. Conversely, with VMware vSphere Foundation (VVF), the complimentary vSAN license available is 100 GiB per core. For instance, if a customer has a 3-node cluster with a total vSAN TiB storage capacity of 100 TiB, and each host has 2 CPUs with 16 cores each, the total VCF calculation for the 3-node cluster would be 3 hosts x 2 CPUs per host x 16 cores per host = 96 cores. Therefore, a customer purchasing a VCF bundle for 96 cores would receive 96 TiB as complimentary vSAN usage. However, it is important to note that in this scenario, the customer is short of 4 TiB. Consequently, according to the new licensing model, the customer would need to pay an additional cost for 4 TiB of vSAN capacity. On the other hand, if a customer has 80 TiB of total vSAN capacity in their cluster, they exceed the vSAN entitlements because, according to VCF, they are entitled to use 96 TiB of vSAN capacity in their cluster. The surplus of 96 TiB – 80 TiB = 16 TiB, meaning the customer does not need to make an additional vSAN capacity purchase. Nonetheless, it is crucial to highlight that there are no refunds or concessions in VCF licenses if you are using less than the vSAN entitlements.

To summarize the changes for vSAN.

- The vSAN subscription capacity is based on the total raw physical storage (TiBs) claimed by vSAN on all the ESXi hosts in each vSAN cluster associated with the vCenter Server or VCF instances that customers plan to license for vSAN.

- vSAN subscription capacity = Total number of TiBs claimed by vSAN in each ESXi host × number of ESXi hosts in each cluster.

Here is the Broadcom Knowledge Base article explaining more on the calculations for different bundles of products: *https://knowledge. broadcom.com/external/article/313548/counting-cores-for-vmware-cloud-foundati.html*

Also, Broadcom has created a PowerCLI tool to identify the number of Core licenses (with a minimum of 16 cores per physical CPU) and TiB licenses that are required to properly license the following VMware products: VMware Cloud Foundation (VCF), VMware vSphere Foundation (VVF), and VMware vSAN for current environments.

The tool is based on the below general licensing notes:

- Each core requires a single license. These licenses do not come in bundles of 16.

- 16 cores are the minimum requirement to be purchased per CPU/physical processor.

- Each TiB claimed by vSAN requires a single license. There are no license minimums.

Let us now look further into the Power CLI calculation script that helps you determine the core count you would need for your Broadcom-VMware licenses upcoming renewals. Understanding this script is important for you to make informed decisions and help with your negotiations.

The customer needs to provide the following inputs for this PowerCLI tool.

1. CLUSTER_NAME: This column displays the name of the cluster. It is an input for providing mapping of core counts for your cluster.

2. NUMBER_OF_HOSTS: This columnspecifies the number of hosts in that cluster.

3. NUMBER_OF_CPU_SOCKETS: This column specifies the number of CPU sockets in that respective physical ESXi host.

4. NUMBER_OF_CPU_CORES_PER_SOCKET: This column specifies the number of CPU cores you have in the physical CPU socket of the ESXi host.

5. VSAN_ENABLED_CLUSTER: This column specifies if this particular cluster is vSAN enabled.

6. TOTAL_RAW_VSAN_TIB: This column is required if VSAN is enabled, then what is the total RAW VSAN capacity of the cluster in TiB.

Table 2-2 is the sample.csv with a variety of inputs for your better understanding.

Table 2-2. *PowerCLI tool inputs to calculate the core requirements with VCF or VVF bundles*

CLUSTER_ NAME	NUMBER_ OF_HOSTS	NUMBER_ OF_CPU_ SOCKETS	NUMBER_OF_ CPU_CORES_ PER_SOCKET	VSAN_ ENABLED_ CLUSTER	TOTAL_ RAW_ VSAN_TIB
Cluster 1	2	2	10	No	0
Cluster 2_VSAN	3	2	20	Yes	100
Cluster 3_VSAN	3	2	20	Yes	150
Cluster 4	2	2	12	No	0
Cluster 5	2	2	16	No	0

Download the script from the following Broadcom site and extract it on your local computer: *https://knowledge.broadcom.com/external/ article/312202/license-calculator-for-vmware-cloud-foun.html*

Figure 2-1 shows a sample calculation based on the core count in your cluster along with vSAN considerations for your cluster.

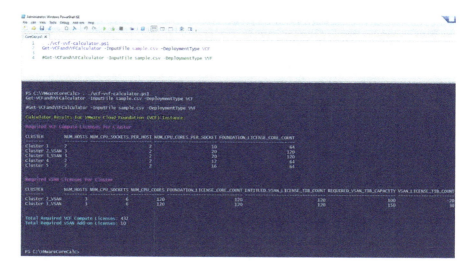

Figure 2-1. *VCF core calculation sample*

Looking further into Figure 2-1, the following are the inferences.

1. For Cluster 1, we have 2 hosts in the cluster and 2 physical CPUs per server with 10 sockets per CPU. The total number of actual cores is 40. However, based on the Broadcom license minimum, 16 core is required per CPU socket. So the total number of cores a customer would need to purchase for Cluster 1 is 64 cores.

2. For Cluster 2_VSAN, the cluster has three nodes, and each node has two CPU sockets. Per CPU socket has 20 cores. Since the minimum number is 16 cores per socket, here we already have 20 cores per socket, so for the core calculations, number 20 is the basis for total core calculations. The total core requirements as per Broadcom licensing for this cluster is 120 cores. Also, with regards to vSAN, if you look at the table below, since as part of VCF, 1 TiB/Core is

     complementary licenses available per core, 120 TiB
     licenses are the total vSAN entitlements available
     in this cluster. Since the total vSAN capacity already
     present is 100 TiB, 20 TiB is the surplus capacity
     from this cluster.

3.   Cluster 3_VSAN is like Cluster 2_VSAN. The only
     difference is that the total vSAN capacity present
     in the cluster is 150 TiB. Since the complementary
     vSAN entitlement based on VCF calculations is
     120 TiB, the total deficiency is 30 TiB of vSAN
     entitlements for this cluster.

4.   Cluster 4 and Cluster 5 are again sample
     calculations based on the core count on each
     CPU socket.

As a final output, the script is summarizing the total VCF core requirements for the environment as 432. The net vSAN add-on capacity needed is 10 TiB.

Please also note the VVF core calculations are also the same, except for the fact that with VVF, the complementary vSAN capacity available is 100 GiB/Core. The PowerCLI script has the option to calculate VVF estimations as well.

2.1.2.3 Cost Increase and Longer Contracts

Because of the bundling of different individual products, the overall price of the combined bundle is higher. On top, existing VMware customers do not have any choice but to move to the new bundle subscription offering if they want to keep their VMware product technology with their deployments. The net cost impact for any deployment depends upon what products you are currently using and how many cores you are running in your environment. Broadcom is providing commitment options for 1 year,

3 years, and 5 years. Broadcom claims that the longer the commitment, the higher the discounts. In addition, Broadcom provides options for the yearly payments for longer commitments as well, where customers can lock the prices per core whey they sign up for longer commitments (3 years or 5 years).

Industry resources suggest that based on the VMware product usage, the renewal cost for many customers has gone up significantly. Where some customers reported an increase from 1.5x to some customers reporting a 5x increase.

To summarize, Broadcom's acquisition has brought considerable changes to VMware's licensing models, product bundles, and overall strategy. While Broadcom continues to bring in more changes even after several months of the acquisition, based on Broadcom's past acquisitions and business practices, VMware changes are following a similar trend, and below are some key points from that trend.

1. **Rationalization of product offerings:** Broadcom usually focuses on optimizing the product portfolio of their acquisitions, streamlining offerings, and discontinuing redundant or underperforming products. Consequently, VMware has also seen restructuring of its product line, impacting licensing models overall.

2. **License consolidation:** Broadcom has previously integrated the licensing models of its acquired companies, enabling a more unified approach. Consequently, VMware also adopted a more unified licensing model across its product suite.

3. **Increased focus on subscription:** Broadcom has been shifting toward a subscription-based revenue model, and VMware has followed suit. By pushing

customers to a subscription-based revenue model, Broadcom aims to keep a constant revenue stream that will possibly uplift Broadcom's dominance in the industry.

2.2 Deciphering New Product Bundles

In this section, we are going to explain more on the different bundle offerings; talk in detail about the core features of different bundles like VCF, VVF, the newly introduced VCF Edge, the vSphere Standard, and the vSphere Essential Plus bundles; and also discuss the specific "add-ons" that can be purchased separately with specific core bundles.

2.2.1 VMware Cloud Foundation (VCF)

VMware Cloud Foundation (VCF) stands out as a premier software package featured prominently on Broadcom's marketing roster. VCF integrates VMware's key product suite to enable customizable and streamlined private cloud solutions for clients. The essential VMware components included in VCF are vSphere (Compute), vSAN (Storage), NSX (Networking), and Aria (Management), all consolidated into one comprehensive package. Essentially, VCF serves as a comprehensive Infrastructure as a Service (IaaS) solution, providing software-defined compute, networking, security, and management capabilities. Here is a list of the available features of VCF:

- The platform offers a comprehensive self-service infrastructure solution for deploying VMs and containers, thereby improving developers' flexibility.

- VMware's robust platform includes essential elements like vSphere, NSX, and vSAN, providing built-in resilience, scalability, and clustering to ensure continuous operations.

- The automation toolset enables infrastructure scaling without the need to increase staff, thus delivering cloud consumption models on-premises. Additionally, the robust automation and orchestration features simplify Day 0, Day 1, and Day 2 tasks.

VMware Cloud Foundation attempts to provide a streamlined and automated solution, streamlining the implementation of a comprehensive infrastructure-as-a-service (IaaS) stack. By promoting uniform, secure, and adaptable operations in both private and public cloud environments, VMware Cloud Foundation promises that businesses can easily expand their infrastructure to accommodate changing demands. By combining multiple features into a single, cohesive platform, VMware Cloud Foundation may reduce the complexity and extra expenses associated with private cloud configurations. Having understood the key features offered with the VCF bundle, now let us look at the components in detail.

2.2.1.1 VMware SDDC Manager

VMware SDDC (software-defined data center) Manager serves as a robust and all-encompassing cloud management platform, offering organizations a centralized hub for controlling their complete SDDC infrastructure. This essential component of VMware's SDDC framework, alongside vSphere, NSX, and vSAN, streamlines the setup, customization, and oversight of VMware's SDDC infrastructure, enabling organizations to efficiently handle their virtualized environments and leverage the perks of software-defined infrastructure. The SDDC Manager offers features and capabilities to help organizations automate and streamline their SDDC operations. For instance, it allows for the deployment and configuration of VMware's NSX network virtualization platform and vSAN, VMware's software-defined storage solution. Additionally, the SDDC Manager provides advanced analytics and monitoring capabilities, empowering organizations to gain a deeper understanding of their SDDC infrastructure and proactively address any potential issues. The SDDC Manager offers advantages through its capability to support policy-based automation. By utilizing the SDDC Manager, companies can establish policies that regulate the setup and performance of their SDDC infrastructure. These policies can then be implemented automatically throughout the entire environment. This feature aids in maintaining consistent configuration of the SDDC infrastructure, aligning with industry best practices. Ultimately, this can enhance security measures and minimize the likelihood of errors and downtime. The SDDC Manager offers assistance for multi-tenancy and resource pooling, allowing organizations to establish separate environments for various applications and workloads. This feature enhances security, minimizes resource conflicts, and simplifies resource allocation and management for different teams and projects.

In general, VMware SDDC Manager is a robust and flexible cloud management platform that offers companies a centralized control point for their SDDC infrastructure. It enables policy-based automation, multi-tenancy, and resource pooling, streamlining the deployment, setup, and administration of VMware's SDDC infrastructure. Through the utilization of SDDC Manager, companies can enhance the effectiveness, flexibility, and security of their SDDC environments, maximizing the advantages of software-defined infrastructure.

2.2.1.2 VMware vSphere

VMware vSphere is a complete virtualization platform that helps businesses manage and optimize their IT infrastructure. At its heart, vSphere provides a framework for operating and controlling virtual machines (VMs) on real servers, therefore consolidating resources and increasing utilization. This virtualization engine takes advantage of hypervisors, which act as a bridge between actual hardware and virtual machines, allowing various operating systems and applications to run concurrently on a single server. vSphere goes beyond mere virtualization to provide a range of tools for managing and automating virtual environments. With features like vCenter Server, which makes it simple for administrators to provision, monitor, and manage virtual machines (VMs), storage, and networking, they can have centralized control over their infrastructure. The platform's comprehensive security features protect against threats and maintain data integrity, while vMotion and DRS provide easy migration and load balancing. Furthermore, vSphere's integration with VMware's broader portfolio of solutions, such as vSAN, NSX, and vRealize, expands its capability and creates a holistic ecosystem for operating modern data centers.

2.2.1.3 VMware vSAN/vVOLs

VMware vSAN and vVOLs are two essential tools used in modern data center virtualization. vSAN is a software-defined storage solution that virtualizes physical storage resources, allowing for the establishment of highly available and scalable storage pools. It eliminates the need for traditional storage arrays while providing a versatile and dependable storage infrastructure for virtual machines. vVOLs, on the other hand, is a virtualization technology that separates the storage layer from the virtual machines, allowing each VM to have its own dedicated storage volume. This results in increased isolation, faster performance, and simpler storage management. By integrating vSAN and vVOLs, companies can create a highly efficient and scalable storage infrastructure that meets the demanding needs of current virtualized environments. This combination has various advantages, including improved storage performance and reliability, easier management, and better security and compliance.

2.2.1.4 VMware NSX

A network virtualization framework called VMware NSX makes it possible to create whole networks in software that closely resemble the conventional physical network infrastructure. It makes it possible to completely divide the virtual network from the underlying physical network, resulting in a network infrastructure that is more adaptable, safe, and effective. To build and administer virtual networks, including logical switches, routers, firewalls, and load balancers, NSX makes use of a centralized control plane called the NSX Manager. A distributed virtual switch, installed on each ESXi server in the network, is then used to spread these virtual network resources throughout the underlying physical network.

NSX provides several benefits, including:

- **Agility**: NSX enables network administrators to quickly and easily provision, modify, and decommission virtual networks, reducing the time and effort required to make changes to the network. This agility is achieved using software-defined networking (SDN) principles, which allow for the decoupling of the network control plane from the data plane.

- **Security:** Micro-segmentation, one of the security capabilities offered by NSX, allows the development of fine-grained security policies that can be applied to specific virtual machines (VMs) or applications. This enhances the network's overall security posture and lowers its attack surface. A distributed firewall that provides stateful inspection and virtual machine-level firewalling is also included with NSX.

- **Efficiency:** By enabling several virtual networks to operate on a single physical network infrastructure, NSX lowers network overhead. This makes fewer physical network devices necessary and makes network management easier. The construction of virtual Layer 2 networks over Layer 3 boundaries is made possible by technologies like Virtual Extensible LAN (VXLAN).

- **Scalability**: Because of its scalability, NSX can operate in expansive, intricate network environments. It supports up to 16 million virtual networks with a maximum of 8,000 ESXi hosts per NSX Manager. Network administrators can design, administer, and secure virtual networks in software with NSX, a potent network virtualization technology. Because of its scalability, efficiency, security, and agility, it is the best option for contemporary network infrastructures.

2.2.1.5 VMware Tanzu Kubernetes Grid

VMware Tanzu is a suite of offerings intended to assist enterprises in developing, executing, and overseeing software across various cloud environments. Its main goal is to make modern applications easier to develop and run by utilizing contemporary architectures like microservices and containers along with contemporary development methodologies like DevOps and GitOps. The Tanzu Kubernetes Grid, an open-source, fully conformant distribution of Kubernetes that offers a reliable, upstream-compatible base for developing and executing contemporary applications, is the centerpiece of the Tanzu offering. Tanzu Kubernetes Grid offers a uniform platform for application development and deployment by making it simple for enterprises to establish and manage Kubernetes clusters across various clouds and infrastructure providers. A full-stack, compliant Kubernetes platform, VMware

Tanzu Kubernetes Grid assists businesses in managing and executing containers at scale. Its goal is to make Kubernetes cluster deployment and maintenance easier in a variety of settings, such as edge locations, public clouds, and on-premises data centers. Tanzu Kubernetes Grid offers an easy-to-deploy, manage, and upgrade Kubernetes infrastructure that is consistent and compliant with upstream releases. Compatibility with a broad range of tools and applications is ensured by its full conformance with the Kubernetes API and its construction on top of the open-source Kubernetes project.

Tanzu Kubernetes Grid is a platform for containerized applications because of several important features. Its compatibility with many contexts, including public clouds, edge locations, and on-premises data centers, is among the most significant. This enables businesses to install and maintain Kubernetes clusters wherever it best suits their requirements, be it in the cloud or in their own data centers. In addition, Tanzu Kubernetes Grid has several capabilities, including intelligent resource allocation, centralized management, and automatic cluster provisioning that make it easier to set up and maintain Kubernetes clusters. Tanzu Kubernetes Grid's support for Kubernetes distributions that are compliant with upstream updates is another important feature. As a result, businesses may seamlessly migrate to a container-based infrastructure by using the same Kubernetes releases and technologies that they are already accustomed to. Along with supporting the Kubernetes ecosystem, Tanzu Kubernetes Grid offers tools like Helm, Flux, and Istio, making it simple for businesses to launch and maintain a variety of containerized applications. Tanzu Kubernetes Grid has capabilities to guarantee the security and compliance of containerized workloads, in addition to supporting various environments and upstream-compatible Kubernetes deployments. By supporting network security policies, image scanning, and automated compliance checks, among other features, these allow businesses to swiftly and simply implement security and compliance guidelines throughout their whole container architecture.

Overall, VMware Tanzu Kubernetes Grid is an effective tool for businesses wishing to scale up the deployment and management of containerized workloads. For enterprises wishing to implement a container-based infrastructure, its support for several environments, upstream-compatible Kubernetes releases, and an extensive array of security and compliance capabilities make it the perfect option.

2.2.1.6 VMware Aria

Private, public, and hybrid cloud systems may all be managed and optimized by enterprises using VMware Aria Suite, a comprehensive platform for VMware solution administration. The suite is made up of multiple products that work together to provide services including cost management, cloud optimization, performance management, and infrastructure automation.

With the help of VMware Aria Automation, formerly known as vRealize Automation, businesses can reduce labor costs and increase productivity by automating the provisioning and administration of infrastructure and applications. It offers users a self-service portal so they can easily request and manage their resources. Additionally, it interfaces with other VMware and outside products, giving consumers a smooth experience.

With the help of VMware Aria Cost Management, formerly known as vRealize Cost, businesses can optimize their expenditure and cut down on waste by getting comprehensive insights into their cloud costs. It offers a thorough overview of cloud expenses, including billing, use information, and chargebacks. Additionally, it provides suggestions for cost optimization, assisting businesses in making cost-effective use of their cloud resources.

For cloud settings, VMware Aria Operations (previously known as vRealize Operations) offers performance management and monitoring services. It gives businesses real-time visibility into the functionality and condition of cloud apps and infrastructure, empowering them to proactively identify and fix problems. Additionally, it provides predictive analytics, which aids firms in anticipating possible problems before they arise.

Network visibility and security services are offered by VMware Aria Network Insight, formerly known as vRealize Network Insight, for cloud settings. It gives enterprises real-time visibility into traffic patterns, network topology, and security posture, empowering them to swiftly identify and address security issues. Additionally, it provides micro-segmentation services, which aid in enhancing an organization's security posture.

In conclusion, VMware Aria Suite is an all-inclusive platform for cloud management that helps businesses optimize and manage their cloud infrastructures. Its products assist businesses in cutting costs, increasing efficiency, and strengthening security by offering automation, cost management, performance management, and security services. VMware Aria Suite gives users a smooth experience and makes it easy for them to manage their cloud resources thanks to its interconnected solutions.

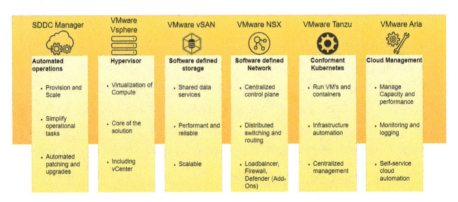

Figure 2-2. *VMware Cloud Foundation components*

In addition to the key features offered with VMware Cloud Foundation, there are a list of additional services users can add as a top-up on the VCF features. Table 2-3 provides the complete list of these key features and add-on services.

Table 2-3. *VMware Cloud Foundation key features and add-on services*

Vmware Cloud Foundation	Services Add-Ons	Use-Case Specific Add-ons
SDDC Manager	Technical Adoption Manager	Advanced Load Balancer
vCenter	Support Account Manager	Vmware Cloud Disaster Recovery
vSphere	Dedicated Technical Support Engineer	Ransomware Recovery
Tanzu Kubernetes Grid		NSX Firewall with ATP
vSAN (1TiB)		NSX Firewall
NSX Networks		vSAN (Additional Capacity)
HCX		Site Recovery Manager
Data Services Manager		Tanzu Intelligence Serivce
Aria Lifecycle Manager		Tanzu Application Service
Aria Operations		Tanzu Spring Runtime
Aria Ops for Logs		Tanzu Data Service Management
Aria Ops for Networks		Private AI Foundation
Aria Automation		
Select Support options		

Please note the add-ons are a separate feature that customers would need to buy separately along with the VMware Cloud Foundation bundle.

2.2.2 VMware vSphere Foundation (VVF)

VMware vSphere Foundation (VVF) is VMware's solution for data center optimization in traditional vSphere environments. It includes Tanzu Kubernetes Grid in addition to Aria Operations and Aria Operations for Logs as part of its standard suite of features. With its intelligent operations, VMware vSphere Foundation is a new offering that optimizes your IT infrastructure. It is the enterprise workload engine that helps businesses of all sizes increase operational effectiveness, boost workload performance, strengthen security, and spur innovation. It provides proactive and predictive operations management that is designed to maximize the efficiency, availability, and performance of your applications and infrastructure. VMware vSphere Foundation enables advanced troubleshooting and delivers intelligent operations management, purpose-built to enable the best performance, availability, and efficiency from your infrastructure while providing unified, holistic visibility and analytics in one place. It has an enterprise-ready IaaS control plane to provision VMs, K8s, and infrastructure services. The list of goods included in the VVF bundle is shown in Figure 2-3.

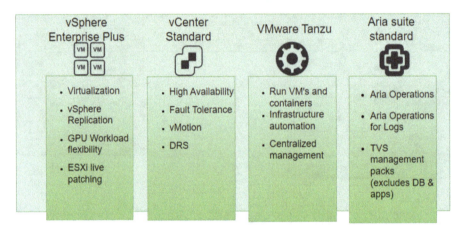

Figure 2-3. *VMware vSphere Foundation components*

Please note that with the VVF bundle as well, customers can add specific add-ons only. The list of add-ons is limited here. If there is a requirement of any add-ons that is not in the list, customers need to work with their Broadcom representatives and ask for it separately. Table 2-4 presents the different products present in the VVF bundle and use-case-specific add-on services that can be added with the VVF bundle.

Table 2-4. *VMware vSphere foundation key features and add-on services*

Vmware vSphere Foundation (VVF)	Use-Case Specific Add-ons
vCenter Server Standard	vSAN (Additional Capacity)
vSphere Enterprise Plus	Site Recovery Manager
Tanzu Kubernetes Grid	Vmware AVI Load Balancer
vSAN (100 GiB/Core)	
Aria Operations	
Aria Ops for Logs	

It should be noted that the VVF bundle does not offer any extra features that are not included in the VCF bundle. Only a fraction of the VCF bundle is VVF. The VCF bundle also includes all of the features included in the VVF bundle. The list of various feature set comparisons between VCF and VVF is shown in the graphic below. The various feature sets and their corresponding availability in the bundles are shown in Table 2-5. It is evident that VVF is a subset of the VCF bundle.

Table 2-5. *Feature mapping with VCF and VVF bundles*

Features	Previous Products	vSphere Foundation(VVF)	VMware Cloud Foundation(VCF)
Compute	**vSphere**		
Distributed Resource Scheduler, Distributed Switch	vSphere Enterprise Plus	•	•
Cross-VC vMotion, Long Distance vMotion, Direct Path vMotion, Storage vMotion	vSphere Enterprise Plus	•	•
High Availability, Fault Tolerance, Data Protection, Trust Authority	vSphere Enterprise Plus	•	•
Kubernetes Runtime, Automated Multicluster Operations	Tanzu Kubernetes Grid	•	•
vCenter: Backup and Restore, Linked Mode, HA	vCenter Server Standard	•	•
Storage	**vSAN**		
Data-at-rest and Data-In-Transit Encryption	vSAN Enterprise	100 GiB / Core	1TiB / Core
Stretched Cluster with Local Failure Protection	vSAN Enterprise	100 GiB / Core	1TiB / Core
Petabyte Scale, Disaggregated Storage for vSphere	vSAN Enterprise	100 GiB / Core	1TiB / Core
Dedup & Compression	vSAN Enterprise	100 GiB / Core	1TiB / Core
Networking	**NSX**		
Networking: Distributed Switching and Routing	NSX Enterprise Plus		•
Large Scale Workload Migration	HCX Enterprise		•
Network Ops: Flow Analysis, App Discovery, M-Seg Planning, Network Assurance and Verification	Aria Operations for Networks		•
Management	**Aria/vRealize**		
Operations: Performance Optimization, Capacity Management, Compliance, Monitoring and Troubleshooting, Log Analytics	Aria Operations	•	•
Automation: Automated Lifecycle Management, App/Infra Provisioning, Governance	Aria Automation		•
VMware and Third-Party Database, Middleware and Application management packs	Aria Operations Enterprise		•
Monitoring and Troubleshooting for Applications with OpenSource Telegraf	Aria Operations Enterprise		•
Out-of-the-box Monitoring and Troubleshooting for Curated Applications with Telegraf agent	Aria Operations Enterprise		•
Native Public Cloud Monitoring	Aria Operations Enterprise		•
TVS management packs	Aria Ops for Integrations	•Excludes DB & apps	•
SDDC Manager: Workload Domain Management, Lifecycle Management, Certificate Management	SDDC Manager		•
Support			
Production Support – Regional Coverage, 24x7 Support		•	•
Select Support – Global Coverage, 24x7 Support, Faster SLAs, SDK/API Guidance			•
Includes Activation and Upgrade Support (Requires full SDDC Stack Deployment)			•

2.2.3 VMware Cloud Foundation (VCF) Edge

An integrated software platform called VMware Cloud Foundation Edge was created to give businesses the capacity to set up and maintain edge computing environments. By acting as a link between decentralized computing needs at the edge and core data centers, it helps enterprises process data closer to its point of generation, maximizing its value.

2.2.3.1 Key Components

Technically, the VMware Cloud Foundation Edge VCF Edge offering is a VCF bundle with a few limitations. The principal constituents of VCF Edge are as follows.

VMware Infrastructure

The VCF edge solution, similar to VCF, is built on VMware's technologies, including VMware vSphere for compute virtualization, VMware vSAN for storage virtualization, and VMware NSX for network virtualization. This familiar architecture allows organizations to leverage their existing VMware skills, thereby reducing training costs and minimizing deployment timelines.

Hyper-converged Infrastructure (HCI)

VMware Cloud Foundation Edge integrates networking, compute, and storage onto a single software-defined platform through the use of a hyper-converged architecture. Performance is improved, and the complexity usually involved in maintaining disparate hardware components is decreased by this consolidation. It provides VSAN functionality, like a VCF bundle.

- **Automation and orchestration**: The edge platform offers automation capabilities that simplify the deployment and maintenance of edge resources, much like VCF does. Organizations may automate routine processes with integrated orchestration capabilities, which lowers the need for manual intervention and the related operational overhead.

- **Networking:** VCF Edge offers NSX networking components that allow leveraging software-defined networking capabilities at the edge locations.

- **Multi-cloud support:** VMware Cloud Foundation Edge
  is designed with interoperability in mind, supporting
  multiple cloud environments. This feature allows
  businesses to seamlessly extend their applications
  and workloads across edge, private, and public cloud
  infrastructures.

Figure 2-4 is a sample presentation of the VCF Edge bundle.

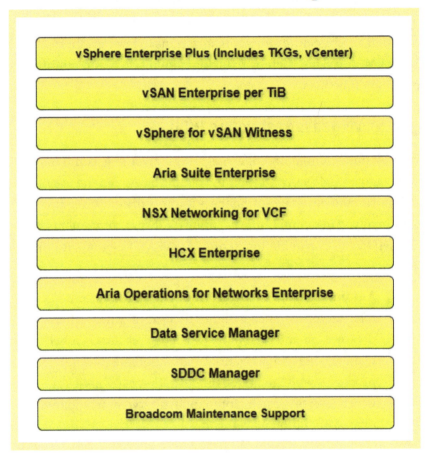

VMware Cloud Foundation Edge

vSphere Enterprise Plus (Includes TKGs, vCenter)

vSAN Enterprise per TiB

vSphere for vSAN Witness

Aria Suite Enterprise

NSX Networking for VCF

HCX Enterprise

Aria Operations for Networks Enterprise

Data Service Manager

SDDC Manager

Broadcom Maintenance Support

Figure 2-4. *VMware Cloud Foundation Edge*

If we look carefully at the figure, then the bundle offering is very similar to VCF. However, the difference is mainly from the licensing perspective. Below are the salient features for the licensing with VCF Edge.

- Subscription-based offering only.

- Per Core pricing. 8 cores per CPU minimum (announced in June 2024); previously it was also 16 cores per CPU minimum.

- Pricing per core per year with a 3-year minimum term (annual billing available). Additionally, if the consumer wants this for just a 1-year period, it will rely on their negotiating power.

- vSAN: 1 TiB per core

- This includes vSphere for vSAN Witness, which allows the vSAN Witness Appliance to be deployed locally. vSphere for vSAN Witness requires vCenter 8.0 U2.

- Requirements:

  - Edge Locations only

  - Minimum of 25 sites required to purchase this product

  - Maximum of 256 cores per vCenter per instance per site (except for manufacturing factory automation, which can have multiple instances per site)

As the adoption of edge computing grows, VMware Cloud Foundation Edge emerges as a viable option for businesses seeking to leverage real-time data processing. Through its range of capabilities, seamless integration with current setups, and various advantages, VMware Cloud Foundation Edge tackles the obstacles presented by edge environments while also opening possibilities for innovation at the edge. Companies that embrace this revolutionary technology will be well-prepared to navigate the intricacies of an interconnected world and propel their digital transformation endeavors.

2.2.4 VMware Vsphere Standard (VVS)

A server virtualization solution that offers improved application availability and data center consolidation is VMware vSphere Standard (VVS). It is made for businesses that don't require the entire range of features and services included in higher-end editions but still want industry-leading virtualization capabilities.

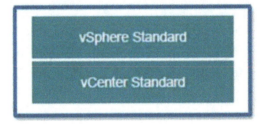

Figure 2-5. *VMware vSphere Standard*

Below are the key features of the licensing with this product.

- Subscription-based model

- Pricing based on per-core usage

- Minimum requirement of 16 cores per CPU

- Flexible terms available: 1-year, 3-year, and 5-year options

- Includes production support

Broadcom is applying restrictions for the use of the VVS bundle, and below are the listed use rights and limitations as suggested on Broadcom site: $https://ftpdocs.broadcom.com/cadocs/0/contentimages/VMware-vSphere-Standard-SPD-Final-May2024.pdf$

"Use Rights and Limitations

- *vCenter Server. vCenter Server may be used to provide centralized management capabilities to any licensed VMware by Broadcom infrastructure environments with an active subscription to Support and Subscription Services.*

- *Restrictions on Use with Public Cloud Services. Customer must not (and must not allow Customer's Third-Party Agents to) use or deploy the Software on any Cloud Services.*

- *Hosting Rights and Restrictions. Customer may use the Software to deliver Internally Developed Applications as a service to a third party via an internal or external network. Except as expressly provided in this paragraph and the License Agreement, the use of the Software for any other types of hosting or for the benefit of any third party in any manner is strictly prohibited unless Customer is an authorized participant in a VMware program that is governed by a separate set of terms and conditions which authorizes such activity."*

An excerpt from the Broadcom website is seen above. Unlike VCF and VVF advertising and use, Broadcom is not aggressively promoting this package as a prospective customer choice, which makes it crucial to present in this chapter. The fact that this is the least expensive product in comparison to the others may be the reason, and users who want to leave VMware soon may wish to use it as a stepping stone.

2.2.5 VMware vSphere Essential Plus (VVEP)

A complete virtualization solution designed for small enterprises with 20 or fewer servers is VMware vSphere Essentials Plus (VVEP). With the help of an all-in-one solution, small businesses may safeguard their operations, streamline administration, and reduce IT expenses by utilizing an affordable virtualization platform. It's the perfect option for small enterprises or academic users who want to virtualize their physical servers in order to save money on hardware while maintaining excellent application availability.

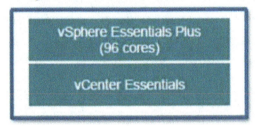

Figure 2-6. *vSphere Essential Plus*

Below are the key features of the licensing with this product.

- Subscription-based licensing model

- Offers a starter pack with 96 cores

- Stackable: 2×96-core option available, totaling 192 cores

- Minimum requirement of 16 cores per CPU; maximum of 3 hosts

- Flexible terms available: 1-year, 3-year, and 5-year options

- Includes production support

The features available with VVS and VVEP are very limited. Before considering this as an option for your deployments, it is important to understand the different features available to you for your use. Let's take a closer look (Table 2-6) at the different feature comparisons between VVEP, VVS, and VVF.

Table 2-6. *Comparison between VVEP, VVS, and VVF*
Source: https://www.vmware.com/content/dam/ digitalmarketing/vmware/en/pdf/docs/vmw-datasheet-vsphere- product-line-comparison.pdf

	vSphere Essential Plus Kit (VVEP)	vSphere Standard (VVS)	vSphere Foundation (VVF)
Licensing Model	Subscription	Subscription	Subscription
License Metric	Core-based (kit includes 96 core licenses)	Per core	Per core
vCenter Edition	Includes vCenter Essentials	Includes vCenter Standard	Includes vCenter Standard
Aria Suite Standard	No	No	Yes
VMware Tanzu® Kubernetes Grid™ Service for vSphere	No	No	Yes
vSAN Enterprise (100 GiB per core)	No	No	Yes, requires 8.0U2B

Let's take a closer look at the different key features comparison. Please note the differences, especially with sections "Simplified Operations," "Built-in Security," "Application Performance," and "Business Continuity." Table 2-7 helps customers to provide inputs on different feature sets available that help them decide on their options based on the respective business needs. If Broadcom is willing to sell you these bundles and if you like to intelligently mix with critical and non-critical workloads based on the business needs, then selecting VVS or VVEP can drastically reduce your overall cost for Broadcom renewals.

Table 2-7. *Different feature sets with VVEP, VVS, and VVF bundle*

	vSphere Essential Plus Kit (VVEP)	vSphere Standard (VVS)	vSphere Foundation (VVF)
Admin Services and Intelligent Operations Management			
vCenter Lifecycle Management Service	No	Yes	Yes
VMware Aria Operations: Efficient Cost and Capacity Management	No	No	Yes
VMware Aria Operations for Logs	No	No	Yes
VMware Skyline Health	Yes	Yes	Yes
DevOps Services			
Tanzu Kubernetes Grid™ Service	None	None	Yes
Tanzu Integrated Services	None	None	Yes

(*continued*)

Table 2-7. *(continued)*

	vSphere Essential Plus Kit (VVEP)	vSphere Standard (VVS)	vSphere Foundation (VVF)
Simplified Operations			
vSphere Lifecycle Manager	Yes	Yes	Yes
vCenter Server® Profiles	No	Yes	Yes
vCenter Server Update Planner	No	Yes	Yes
Content Library	Yes	Yes	Yes
vSphere Configuration Profiles	No	No	Yes
Distributed Switch™	No	No	Yes
Host Profiles and Auto Deploy™	No	No	Yes
Virtual Volumes™	No	Yes	Yes
Green Metrics	Yes	Yes	Yes
Built-in Security			
Identity Federation	No	Yes	Yes
vSphere Trust Authority	No	No	Yes
Trusted Platform Module Support	Yes	Yes	Yes

(continued)

Table 2-7. (*continued*)

	vSphere Essential Plus Kit (VVEP)	vSphere Standard (VVS)	vSphere Foundation (VVF)
Virtual TPM	Yes	Yes	Yes
FIPS 40-2, 140-3, and Common Criteria Certification	Yes	Yes	Yes
TLS 1.2 and 1.3	Yes	Yes	Yes
Virtual Machine Encryption	No	No	Yes
Application Performance			
Distributed Resource Scheduler (DRS)	No	No	Yes
Storage DRS	No	No	Yes
Distributed Power Management (DPM)	No	No	Yes
Storage Policy-Based Management	No	Yes	Yes
Business Continuity			
vSphere Hypervisor	Yes	Yes	Yes
vMotion	Yes	Yes	Yes
vCenter® Hybrid Linked Mode	No	Yes	Yes
vSMP - Enables virtual machines to have multiple CPUs	Yes	Yes	Yes

(*continued*)

Table 2-7. (*continued*)

	vSphere Essential Plus Kit (VVEP)	vSphere Standard (VVS)	vSphere Foundation (VVF)
High Availability	Yes	Yes	Yes
Storage vMotion	No	Yes	Yes
Fault Tolerance	No	2 vCPU	8 vCPU
vSphere Replication™	Yes	Yes	Yes
vCenter High Availability	No	Yes	Yes
vCenter Backup and Restore	No	Yes	Yes
vCenter Server Appliance™ Migration	No	Yes	Yes

2.3 Understand Subscription Costs

As Broadcom integrates and modifies VMware's offerings, understanding the different Broadcom Bundle costs becomes crucial for enterprises looking to remain competitive and efficient. Organizations must carefully assess their needs, evaluate the potential cost savings of bundled products, and anticipate future expenses linked to subscription models.

Please take note that the price we are listing here is based on our investigation into various customer blogs and communities, particularly those in the United States. These prices are merely estimates and can differ from what you would pay when dealing with a reseller or your Broadcom agent. This is merely meant to give you a general idea of how much each bundle's offers differ in price. These prices can fluctuate, so it's best to consult a reseller or your Broadcom agent for precise pricing.

Table 2-8. *Reference price ranges in USD for different Broadcom – VMware bundles*

Broadcom-VMware Bundle offering	List range price in USD
VMware Cloud Foundation (VCF) – 1-Year Prepaid Commit – Per Core	$300–$400
VMware Cloud Foundation (VCF) – 3-Year Prepaid Commit – Per Core	$800–$1,100
VMware Cloud Foundation (VCF) – 5-Year Prepaid Commit – Per Core	$1,250–$1,750
VMware vSphere Foundation (VVF) – 1-Year Prepaid Commit – Per Core	$165–$200
VMware vSphere Foundation (VVF) – 3-Year Prepaid Commit – Per Core	$360–$430
VMware vSphere Foundation (VVF) – 5-Year Prepaid Commit – Per Core	$590–$710
VMware Cloud Foundation Edge (VCF-Edge) – 1-Year Prepaid Commit – Per Core	$210–$300
VMware Cloud Foundation Edge (VCF-Edge) – 3-Year Prepaid Commit – Per Core	$500–$850
VMware Cloud Foundation Edge (VCF-Edge) – 5-Year Prepaid Commit – Per Core	$750–$1,000
VMware vSphere Standard (VVS) – 1-Year Prepaid Commit – Per Core	$60–$90
VMware vSphere Standard (VVS) – 3-Year Prepaid Commit – Per Core	$130–$170
VMware vSphere Standard (VVS) – 5-Year Prepaid Commit – Per Core	$230–$270

(continued)

Table 2-8. (*continued*)

Broadcom-VMware Bundle offering	List range price in USD
VMware vSphere Essentials Plus (VVEP) – 1-Year Prepaid Commit – Per Core	$4,300–$5,000
VMware vSphere Essentials Plus (VVEP) – 3-Year Prepaid Commit – Per Core	$9,700–$11,500
VMware vSphere Essentials Plus (VVEP) – 5-Year Prepaid Commit – Per Core	$16,000–$17,200

As the enterprise technology market continues to evolve, businesses that remain agile and informed will be better positioned to reap the benefits while minimizing costs associated with virtualization and cloud infrastructure. By staying informed and strategic, organizations can optimize their investments in Broadcom-VMware product deployments with their infrastructure.

2.4 Analyzing Implications for Existing Contracts and Agreements

Let's examine the ramifications of these licensing adjustments in detail and have a better understanding of the ground dynamics of these changes in this part.

2.4.1 Implications for Existing Perpetual Customers

As per the industry definition, a perpetual license is "an on-time payment that gives user the right to use the particular version of software application indefinitely." With perpetual licensing, a customer purchasing software pays an upfront charge for the software license. They also get supplemental support for a limited period, during which additional benefits are included. After the support period ends, the customer has the option of using the current version of the software without additional support, paying a lower-cost fee to get support and updates, or buying a new license for a new version of the software. Along with the perpetual license, the vendor typically provides technical support for a period of 1–3 years. During this initial period, the vendor usually provides software updates. However, updates might or might not be provided for free in perpetuity. Similar circumstances applied to VMware, where clients had to pay a hefty upfront fee to acquire licenses that gave them everlasting ownership. They then routinely pay less than what would be required up front to receive VMware support and services.

Therefore, considering the nature of perpetual licenses, those who own VMware perpetual licenses today are immune for the period they have an active support contract. Broadcom announcements confirm there is no immediate change for businesses with "active perpetual licenses." However, please note that transitioning to a subscription-based model will be necessary to receive continued support once the existing support contract expires. Also, please note that, if a customer is looking to move out of their VMware deployments, they can continue having their current deployments "as-is" as they are the legal owner of these perpetual licenses. The downside is that Broadcom-VMware would not be providing support for any issues they may want to raise with support teams to get assistance. But on the positive side, Broadcom also announced that they are going to provide "critical only" security patches for customers using vSphere (7.x and 8.x) perpetual licenses even with expired service contracts. Here is the knowledge article from Broadcom that confirms this: *https://knowledge. broadcom.com/external/article?legacyId=97805*

Situation with Expired Perpetual Licenses

Broadcom also announced on April 23, 2024, that they are transitioning from VMware Customer Connect portal to the the Broadcom Support portal, which will migrate all license keys from the VMware portal into its own software management portal. The implication of this movement is that this movement is only bringing over the licenses and contracts that are still active support contracts. Where Broadcom announced that "after **approximately Sunday, May 5, 2024, at 7:30 p.m. (PDT)**, if you have license keys associated with expired support contracts, you won't have access to them in the Broadcom systems."

Customers should check their license inventories and contract details on the new Broadcom portal to ensure that all information has been transferred accurately and completely. This proactive approach will mitigate any potential issues arising from the transition and ensure compliance with the new license conditions.

Link to the announcement: *https://knowledge.broadcom.com/ external/article?articleNumber=282163*

This deletion of old license history has several effects on many stakeholders. First, in the event that you need to reinstall the program or possibly move to a new host, you won't be able to reactivate it if you don't maintain your own records of licenses with expired support. Second, an audit may run into problems if Broadcom no longer has your license documents on file. Furthermore, a flat layout will replace the folder structures that infrastructure teams used to frequently make on the VMware portal in order to maintain track of which keys are assigned to which server farms and clusters. Moreover, any notes pertaining to license keys will be misplaced as well post the move from the VMware customer portal to the Broadcom portal.

2.4.2 Implications for Existing Subscription Customers

VMware licensing in the past also used to offer term-based licenses, which, in other words, is a subscription-based offering where the validity of the licenses was applicable only for a certain duration. Many of the customers in the industry used to buy this type of license instead of perpetual license offerings. As "term-based" licenses suited best for organizations looking for low-cost options, especially when the project requirements are for a defined time duration only. Also, VMware had certain products that were based on cloud SaaS (software as a service) services, like vSphere+, Aria SaaS, NSX+, VMC on AWS offering, etc. But with the Broadcom acquisition of VMware, all the existing subscription customers, whether the holder of "term-based" licenses or running their deployments on Cloud SaaS service, will also be required to convert their subscriptions to the new product bundles (and applicable add-ons) in the new simplified portfolio offered by Broadcom.

2.4.3 Implications in General

There are a few broad ramifications that we can foresee with the licensing adjustments in addition to the effects on current contracts and agreements.

2.4.3.1 Expect Increased Audits

After acquiring VMware, Broadcom has already stated that they anticipate growing revenue by 100% over the next 3 years. In terms of money, Broadcom wants to increase VMware's sales in 3 years from $4.7 billion to $8.5 billion. Because of VMware's market positioning and the aggressive revenue targets that the company has set, it is predicted that Broadcom will pursue clients and squeeze every last cent out of them. In contrast to how it handled its earlier acquisitions of CA Technologies and Symantec, Broadcom has stated that it would conduct the VMware purchase differently. However, based on the whole experience thus far, it appears to be playing by a well-known script with VMware as well.

In the upcoming months, we anticipate that VMware customers will experience an increase in audits. Additionally, users should anticipate that Broadcom will be using aggressive software auditing techniques.

2.4.3.2 VMware's Outreach

A representative from VMware will likely get in touch with you to start a conversation and seek chances to sign clients up for 3-year to 5-year lengthy contracts. It is anticipated that the vendor will lock clients in longer periods and pull maximum business from the product offerings and marketing strategies. Broadcom is aware that they have captive customers and that it may take them 1–2 years to transition away from VMware server virtualization. Thus, Broadcom would want to use any opportunity to increase their revenue because, if they bide their time and fail to turn a profit within the next 1–3 years, their market appeal will suffer regardless. Therefore, the top priority for the Broadcom VMware sales staff would be reaching out to potential customers and selling maximum possible products to continue maintaining dependency on the VMware product portfolio.

2.4.3.3 Renewal Engagement

Considering the shift in dynamics with respect to cost and license model, as increasingly reported by many customers, Broadcom VMware is engaging with customers only nearing their contract expiration, leaving less room for them to decide on the alternative roadmap. For VMware customers, the stakes are high. VMware's suite of products and services is deeply embedded in enterprise IT infrastructures worldwide. With Broadcom's acquisition, customers are anxious to understand how their current contracts will be impacted, how pricing may change, and what new features or capabilities they can expect. The delay in providing this information has caused disruptions in operational planning for many organizations that rely on VMware products.

It is getting increasingly clear that Broadcom is employing the delay tactics to create more pressure on the customers where customers are left with little or no choice and are more likely to go ahead and purchase longer-term subscriptions for VMware products.

2.5 Strategies for Negotiation with Broadcom

Because of the dramatic change in the software market caused by Broadcom's acquisition of VMware, businesses now need to adjust to new licensing requirements and standards. Businesses trying to manage the complexity of Broadcom-VMware licensing must develop strong negotiating techniques. Here are some important tactics to improve the way you negotiate.

2.5.1 Understand Licensing Models

Before entering negotiations with Broadcom, it is crucial to have a deep understanding of both Broadcom's and VMware's new licensing models. Familiarize yourself with

- **Types of licenses:** Identify the different types of bundles and licenses offered (subscription, enterprise agreements) and their implications for costs and usage.

- **Usage rights:** Clarify the usage rights associated with each licensing option, including geographical limitations and user caps.

- **Support and maintenance:** Explore the options for support and maintenance included in the licensing agreement, as these can affect the total cost of ownership and operational efficiency.

Customers would be better equipped to decide on their exact needs if they have a thorough understanding of the new licensing model and the various bundles that are offered for purchase. In contrast to a VCF bundle with add-ons, customers may feel that a VVF bundle with certain add-ons better suits their needs. It is therefore recommended that you comprehend the circumstances completely and are well-versed in the new license model and all of the available options.

2.5.2 Assess Your Position and Needs

Evaluate your organization's current and future needs regarding VMware products. Conduct a comprehensive analysis of your:

- **Current utilization**: Assess your existing VMware deployments and how they align with your business objectives.

- **Project growth or cut down:** Consider anticipated growth or possible shrinkage of VMware footprints over the next few years based on your business needs, as this will influence your overall licensing requirements.

- **Budget constraints:** Establish a clear budget for licensing, which will help you identify viable options during negotiations.

Customers can calculate the core counts required for renewals using the script method highlighted in this chapter as well. The customers can create a detailed report outlining their organization's software usage, projected requirements, and budget parameters. This foundational document will support informed discussions with Broadcom-VMware representatives.

2.5.3 Leverage Existing Relationships

Make use of the connections your company has already established with companies like VMware, Broadcom, or any other reseller of VMware software licenses. Well in advance of the dates of your renewal, you can get in touch with them to obtain the budgetary figures based on market price. Furthermore, preexisting alliances might offer legitimacy and open doors to better conditions.

- **Activity history:** If you've been a loyal customer, emphasize your long-standing relationship and highlight your contributions as a partner to stand a better chance with the negotiations.

- **Feedback on products:** Share honest feedback on prior usage to help tailor the licensing negotiation to better fit your needs.

Using the existing relationship can provide better opportunities to lead the negotiations in your favor. Prepare a summary of past interactions and successful projects using VMware products, showcasing the possible business avenues to gain better discounts and negotiated prices.

2.5.4 Explore Bundling or Trade-In Opportunities

Broadcom often offers bundled licensing packages and existing perpetual licenses "trade-in" options that may present substantial cost savings and simplified management. If you are a customer that uses different Broadcom products, discussing bundling options or trade-in options can yield beneficial agreements that suit your needs.

- **Cross-products discounts**: Investigate potential discounts for bundling VMware products with other Broadcom offerings.

- **Enterprise agreements**: Inquire about enterprise licensing agreements that can provide comprehensive access to multiple products at a reduced rate.

- **Leverage trade-in options:** Consider the benefits of trading in perpetual licenses for subscription licenses. Broadcom offers pricing incentives for this transition, which can be an effective way to update and streamline VMware usage.

In order to find potential savings opportunities, customers can research various bundles and compare them against your single license needs.

2.5.5 Negotiate Flexibility

Licensing agreements must be flexible in the continually evolving technology context. The usage of your environment can be very dynamic in nature. Advocate for clauses that allow scaling up or down, as well as accommodating changes in your organization's structure. For an example, there might be a clause that Broadcom probably would want to impose like a minimum purchase of ~50K USD to allow any add-in or scope increase in between the renewal dates, etc. And not only the scope increase but customers should also have a strategy of scope decrease in between the renewal dates and the possible leverage for scope decrease they can demand from Broadcom. Therefore, it is advised that

- **Revisability clauses:** Seek terms that allow periodic reviews and adjustments based on usage changes or market conditions.

- **Trial periods**: Negotiate trial periods for new products or updates to assess their fit before committing to full licenses.

Frame negotiation points focused on flexibility that align with your organization's growth trajectory and operational needs.

2.5.6 Get Everything in Writing

Once negotiations lead to a workable agreement, ensure every detail is captured in writing. Written agreements mitigate misunderstandings and provide a clear reference point for future engagements.

- **Terms and conditions**: Review the terms meticulously before signing to ensure they align with verbal agreements.

- **Recurring costs:** Make sure to understand the implications of renewal costs and any potential price escalation clauses.

Considering the market dynamics, it is advised to draft a formal checklist of all agreed terms to be included in the official licensing document, ensuring clarity and accountability.

Negotiating licensing agreements with Broadcom-VMware can be a complex process, but with the right strategies in place, organizations can secure favorable terms that align with their operational goals. A thorough understanding of the licensing landscape, combined with a detailed assessment of your needs, strong relationship leverage, exploration of bundling options, advocacy for flexibility, and meticulous documentation, will empower you during negotiations. By adopting these strategies, your organization can navigate the licensing landscape successfully, ensuring continued innovation and efficacy in technology utilization.

2.6 Summary

In this chapter, we discussed VMware's long-standing perpetual license model, on which many partners and customers had depended, has ended with the swift and significant licensing changes brought about by Broadcom's acquisition of VMware. Although the present licenses will continue to work, VMware users will need to consider the implications, weigh their options, and probably start budgeting for a gradual shift to subscription pricing. We discussed on different bundle products, namely, VCF, VVF, VVS, and VVEP, and understood the different add-ons that are available with the respective bundles. We also discussed the cost comparison between these bundles. We discussed the different implications of these license changes on VMware users. The following are the key things to keep in mind:

- VMware perpetual licenses are no longer available for purchase.

- Existing perpetual licenses will eventually lose access to support and updates. But if customers are ok with that, they can continue using their perpetual licenses.

- Broadcom will provide tools and incentives to help transition customers to subscription.

- Alternatives to VMware are available but require careful assessment.

In the end, we discussed some of the negotiation tactics that customers may employ that help them to lead the negotiations with Broadcom.

CHAPTER 3

Navigating Challenges in Transitioning

In the preceding chapter, we thoroughly examined the implications of Broadcom's acquisition of VMware and outlined different bundle products that are available. We also discussed various negotiation strategies pertinent to your forthcoming renewals. This chapter commences with an exploration of the challenges associated with a subscription-based model, as well as the collaboration and responsiveness issues that many organizations encounter when engaging with Broadcom. Additionally, this chapter will delineate the potential evaluation criteria to consider when seeking alternatives to VMware. We will also explore contemporary use cases for workload options that extend beyond VMware. Finally, we will present a comprehensive vendor landscape along with a detailed feature-by-feature comparison of various vendor platforms available in the server virtualization market, juxtaposed with VMware's technology offerings.

3.1 The Sting of Broadcom's Resistance: Cooperation and Response Issues

In the rapidly changing environment of the technology sector, only a select few companies have succeeded in staying ahead in innovation.

A giant in the semiconductor market, Broadcom's products power a vast

© Sumit Bhatia and Chetan Gabhane 2025
S. Bhatia and C. Gabhane, *Navigating VMware Turmoil in the Broadcom Era*,
https://doi.org/10.1007/979-8-8688-1264-4_3

array of devices in today's world. Nevertheless, as the company broadens its influence through strategic acquisitions and alliances, it encounters heightened scrutiny and criticism regarding its collaborative practices and responses. This section delves into the intricacies of Broadcom's resistance to collaboration, the implications of its decisions, and the challenges it faces in maintaining its competitive edge.

3.1.1 The Nature of Broadcom's Business Model

Broadcom functions within a multifaceted ecosystem characterized by swift technological advancements and evolving customer demands. The company's strategy is centered on the acquisition of complementary technologies, which are subsequently integrated into its existing portfolio. This methodology enables Broadcom to deliver holistic solutions and sustain its competitive edge across various profitable sectors, such as networking, broadband, and wireless communications. In 2022, VMware commanded nearly 97% of the revenue share in the server virtualization market, outpacing other providers in the industry. The combination of VMware's popularity and Broadcom's assertive business strategy presents a significant perspective for future business development.

Nevertheless, this assertive approach brings with it a fundamental conflict. Broadcom's reluctance to engage in full cooperation with other companies, especially smaller entities and startups, has generated apprehension among industry participants. Detractors contend that Broadcom's focus on acquisitions hampers innovation and limits the collaborative ethos that is vital for progress in the technology industry.

3.1.1.1 Subscription-Based Model and Long-Term Contracts

Broadcom, a prominent player in the semiconductor and infrastructure solutions sector, has implemented a subscription-based model that relies on long-term contracts to stabilize its revenue. Recent feedback from

VMware customers indicates that Broadcom is urging clients to enter minimum contracts of 3–5 years when transitioning existing customers to the VMware subscription model. While this approach may offer advantages for Broadcom, it also introduces a range of challenges that stakeholders need to consider. The harsh realities of subscription-based and long-term contracts are as follows:

- **Customer Lock-In**

  With the long-term contracts and subscription model, Broadcom aims to have customer retention. However, the reality is that long-term contracts can trap customers into agreements that may not serve their evolving needs. As technology progresses rapidly, businesses may find themselves locked into a contract that doesn't align with their shift in requirements or budget allocations. Longer contracts, especially with a lot of uncertainty on the VMware product roadmap, are causing a lot of frustration and leading customers to aggressively look for alternatives. Broadcom's aggressive business strategy to enforce customers into longer contracts is not at all helping in the current VMware turmoil that industry is facing. From the industry reviews and feedback, Broadcom is less willing to negotiate and is enforcing customers for longer contracts as the only go-to strategy for the company.

- **Revenue Predictability vs. Profitability**

  Although a subscription model offers companies a steady stream of revenue, it does not necessarily guarantee profitability. Long-term agreements often require substantial initial investments in infrastructure, customer support, and maintenance, with the return on investment (ROI) potentially taking longer than anticipated. Broadcom must navigate the challenge of reconciling the immediate expenses associated with acquiring and servicing customers with the long-term advantages of securing those customers for extended durations. While enforcing longer contracts may initially appear beneficial, it can pose difficulties for both the company and its customers in the long run. The company may fail to achieve the anticipated revenue over time, leading to reduced focus and cost-cutting measures regarding product and technology support, particularly concerning VMware products. This scenario is an unwelcome prospect for VMware customers, yet it remains a harsh reality associated with long-term contracts.

- **Innovation Stagnation**

  A long-term subscription model can breed complacency in innovation. With customers locked into agreements, there might be less urgency to evolve product offerings or improve services. If Broadcom shifts focus from exploring new technological advancements to merely fulfilling the contract terms, it risks falling behind the industry dynamics that are associated with virtualization technology. This would severely impact customers who need constant innovation to meet ever-evolving business challenges. The dynamic tech landscape demands a commitment to continual innovation, which could become compromised under long-term agreements.

- **Market Responsiveness**

Technology changes at lightning speed, and current market demands can shift overnight. Broadcom's model needs to strike a balance between long-term contracts and flexibility. If customers foresee new technologies or competitors emerging, they may feel overly constrained by the lengthy commitments they've made. This dynamic can lead to customer dissatisfaction if Broadcom is unable to pivot quickly in response to changing needs, leaving them vulnerable to competitors ready to swoop in with flexibility as their selling point. Gartner predicts that over time, "virtualization market will fragment." There are higher chances that the existing large VMware customer base would eventually look for better alternatives, leading to overall less attractiveness for VMware products.

- **Customer Support Challenges**

Subscription models typically entail a commitment to continuous customer support; however, extended contracts may present challenges regarding the scalability of customer service. As the customer base grows, the intricacies of addressing diverse needs also increase. Broadcom might experience pressure on its support resources as clients with varying contractual agreements and expectations demand customized solutions. This situation could potentially impact the quality of service, resulting in customer dissatisfaction and increased churn rates during contract renewals.

- **Pricing Pressure**

  With long-term contracts, there is a risk of pricing stagnation. Customers might be hesitant to negotiate due to existing commitments, leading Broadcom to potentially miss out on significant increases in line with market demands. As competitors offer new and flexible pricing structures, Broadcom may have to grapple with its pricing strategy to keep existing customers without alienating new ones or undermining its market position.

While Broadcom's subscription-based model and long-term contracts offer a pathway to stable revenue for the organization itself, the associated problems can pose significant challenges. Customer lock-in, revenue predictability issues, innovation stagnation, market responsiveness, customer support challenges, and pricing pressures represent several obstacles that necessitate meticulous management.

3.1.2 The Cooperation Issues

The phrase "The customer is king" encapsulates the fundamental ethos of modern business philosophy, emphasizing that the satisfaction and experiences of customers should take precedence in all aspects of a company's operations. In a competitive marketplace where choices abound, businesses must recognize that their survival hinges not merely on the quality of their products or services but also on the ability to understand and cater to the evolving needs and preferences of their clientele. This customer-centric approach entails actively listening to feedback, anticipating market trends, and embracing innovation to enhance the overall consumer experience. Companies that genuinely honor the notion of customers as kings prioritize building strong relationships, fostering loyalty, and creating an environment where customers feel valued and respected.

However, it seems with Broadcom, this proven fundamental ethos where it is believed that "the customer is king" does not stand true. The aggressive business model of enforcing customers into "longer-term contracts," increasing price, and missselling of bundles is creating an overall feeling that Broadcom is defining the rules for business and not its customers. It is no longer based on what customers need and want; rather, it is based on what Broadcom is willing to sell. It may sound funny, but it is true that in this modern outset, it is prevailing that Broadcom is defining the rules of the engagement with the VMware customers. Moreover, because of the aggressive business strategy around "take it or leave it," it is adding more difficulties to the current situation, leaving less room for VMware customers to adjust to the new dynamics of the business and leading to excessive dissatisfaction. Broadcom's limited willingness to forge cooperations could result in a fragmented ecosystem, where customers have to juggle multiple vendors for various services, ultimately undermining their experience.

For Broadcom to sustain its position as a leader in the tech industry, it must recognize the value of partnerships and open dialogue. By moving toward a more collaborative approach, Broadcom can not only mitigate risks associated with its current stance but also foster an environment conducive to growth, innovation, and increased customer satisfaction. In an age where collaboration is key to success, embracing it may very well be the smartest strategy for Broadcom's sustained dominance in the technological landscape.

3.1.3 The Response Issues

Broadcom's business model may intentionally limit the frequency of responses to customer inquiries and feedback in order to foster a sense of exclusivity and heighten anticipation for their products or services.

This approach aligns with a well-established marketing strategy where businesses generate urgency through tactics such as countdown timers and scarcity messaging, which can evoke a "fear of missing out" among consumers. As a result, this often leads to predictable purchasing behaviors. By employing these strategies, companies can effectively encourage impulse buying, prompting potential customers to make swift decisions rather than deliberating over their purchases. It is important to examine the response challenges associated with Broadcom, particularly in relation to the VMware business model, as highlighted by various industry experts in recent reports.

- **Delayed Support Response Times**

  A prominent challenge faced by VMware customers following the acquisition has been the increased latency in response times for requests. Numerous organizations rely extensively on VMware's solutions for their IT operations, and any interruption or delay in resolving requests may have a substantial effect on business continuity and annual financial planning. The postponement of response times and the delay in the issuance of renewal quotes may push VMware customers closer to their annual renewal deadlines. This situation intensifies the pressure on customers to accept the renewal costs that Broadcom aims to implement, leaving them with limited options and necessitating their renewal agreement to ensure the continued support of their operations.

- **Inconsistent Communication Channels**

  Following the merger, customers have encountered irregularities in communication. The VMware product resellers and managed service providers that previously facilitated the purchase of VMware products are now subject to the regulations established by Broadcom after its acquisition of VMware. This has resulted in inconsistent communication channels, limiting customers' options for obtaining essential information regarding their engagements and renewal requirements.

The challenges arising from the integration of Broadcom and VMware carry considerable consequences for enterprises that depend on VMware's technologies. Lengthened contract durations, slower response times, and difficulties in collaboration are contributing to growing frustration among IT departments and dissatisfaction among customers. Consequently, organizations need to brace themselves for these obstacles and formulate strategies to alleviate their effects.

3.2 Identifying Hurdles and Top Challenges in Seeking Alternatives

VMware has long been recognized as a leading provider in the field of virtualization. According to a Gartner report, in 2022, VMware commanded nearly 97% of the revenue share in the server virtualization market. Nevertheless, the ever-evolving landscape of technology often prompts businesses to consider alternatives that may align more closely with their developing requirements. This is particularly relevant following Broadcom's acquisition of VMware, which has led many organizations to reevaluate their existing infrastructure solutions. However, the pursuit of alternatives to VMware introduces its own unique challenges. In this section, we will examine some of the primary obstacles that organizations encounter when investigating options outside of VMware (ref. Figure 3-1). As organizations strive to enhance their IT infrastructure and achieve cost efficiencies, a growing number are investigating alternatives to VMware technology. This movement is driven by the need for greater flexibility, enhanced performance, cost reductions, and a preference for a more streamlined, cloud-native approach. However, transitioning away from VMware is not without its difficulties. In this section, we will identify some of the significant challenges that businesses face when seeking alternatives to VMware technology, as well as strategies they can implement to address these challenges effectively.

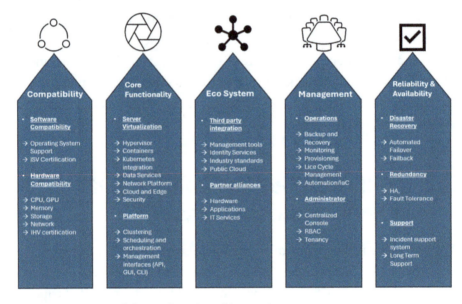

Figure 3-1. *Possible evaluation dimensions*

3.2.1 Understanding Current Dependencies

A major challenge that organizations face when exploring alternatives to VMware is comprehending their current dependencies. Companies that have made substantial investments in VMware products often have a complex array of applications and services that rely on VMware's ecosystem. Therefore, it is crucial to meticulously outline existing dependencies to guarantee that any prospective alternative can sufficiently accommodate current workloads. In this section, we will examine the key dependencies within VMware environments, encompassing hardware specifications, software requirements, networking aspects, and other vital considerations.

3.2.1.1 Hardware Dependencies

The cornerstone of any infrastructure solution lies in its reliance on the underlying hardware. Organizations often encounter challenges related to hardware dependencies when exploring alternatives to VMware. It is crucial to verify that the hardware satisfies both compatibility and performance standards for any viable alternatives that clients may consider. A thorough assessment of hardware requirements should be conducted, along with an inventory tailored to specific business needs. Important hardware components encompass:

- **Server compatibility**: Like VMware products, most of the virtualization vendors have specific server compatibility lists where the platform is certified to run. Organizations must carefully do their hardware inventory and analyze their existing server's remaining life, their compatibility, etc. This would help to explore all possible opportunities to leverage existing hardware when looking for potential alternatives.

- **CPU and memory**: Every virtualization platform requires efficient CPU and memory resources. Servers should have processors that support virtualization (Intel VT-x or AMD-V) and sufficient RAM to accommodate virtual machines (VMs). The ratio of physical RAM to VMs should ensure optimal performance without overcommitting resources. The inventory of all available compute resources allows organizations to weigh their requirements against what they currently have with them already.

- **Storage solutions**: Performance and availability depend on storage solutions. It is required to understand the requirements of storage solutions, including direct attached storage (DAS), network attached storage (NAS), and storage area networks (SAN). Understanding I/O performance and redundancy options is crucial for selecting the right storage solution.

- **Network interfaces**: Adequate network interfaces are essential for virtual machines to communicate effectively. Network cards must support virtualization features and be capable of handling the expected load.

- **Hardware life and warranties**: Apart from identifying the capabilities and physical aspects of the hardware, it is also crucial to know the years of life or EOS (end-of-support) duration remaining on the hardware that is currently present in the environment. This would help in better planning and budgeting when looking for alternatives.

Identifying the primary challenges and obstacles in the search for VMware alternatives necessitates a thorough assessment of the hardware inventory and its capabilities. This evaluation is crucial as it not only provides insight into the financial investment needed for alternative solutions but also aids in developing more effective strategies for transitioning away from VMware when necessary.

3.2.1.2 Software Dependencies

The software environment is crucial for the effective implementation of virtualization solutions. It is important to note that not all software is certified for use with every virtualization solution available in the market. Certain vendor-specific appliances and their associated deployment packages may not be compatible with all virtualization platforms. Additionally, there are legacy applications that may resist modernization or containerization, making the search for alternatives particularly difficult. Essential software dependencies encompass:

- **Operating system**: First and foremost, an inventory of the list of operating systems and their specific versions currently running on the VMware platform is very important to list out. When looking to migrate the workloads to any VMware alternative, the list of operating systems supported comes in handy and helps to easily isolate the unique VMware dependencies you may have in your environment.

- **Licensing**: Numerous products available today offer licensing specifically tailored for VMware platforms. For instance, companies like Oracle and RedHat provide licensing options based on VMware ESXi hosts rather than on a per-guest virtual machine basis. This approach has historically enabled organizations to optimize costs associated with critical software licensing. Additionally, many software vendors exclusively license their products for the VMware platform. When addressing potential challenges, it is essential to meticulously document your dependencies and comprehend the various licensing models (such as per processor or per CPU). This diligence will not only facilitate discussions with the respective vendors regarding the portability of your licenses but also ensure that any new alternatives you consider will keep your workloads compliant, thereby mitigating the risk of software vulnerabilities and associated penalties.

- **Management tools**: Currently, numerous VMware ecosystems are interconnected with various critical infrastructure components. For instance, VMware vCenter facilitates integration with backup solutions, enabling customers to perform "agentless" backups of their virtual machines. Additionally, there are monitoring tools from multiple vendors that provide integration with vCenter or Aria Operations, thereby enhancing the efficiency of their monitoring solutions. Likewise, VMware's management tool portfolio features a range of such integrations. When assessing primary challenges, it is crucial to compile a detailed list of dependencies. This approach allows for a clearer understanding of potential obstacles, ensuring that when exploring alternatives, all necessary requirements are addressed. Consequently, comprehending the configurations and requirements of these management tools is vital for optimizing operational efficiency.

- **Native appliances**: VMware products have been the favored technology for numerous organizations over the years. Consequently, many software vendors have developed their native offerings based on VMware technology solutions. These vendors may possess certified products or may not, and they might also offer various modern alternative virtualization platform technologies. It is essential to catalog these native appliances and assess your business's reliance on them when considering any potential replacements.

Understanding your existing software dependencies is crucial when assessing various virtualization vendors for potential VMware alternatives, as it helps in identifying obstacles and key challenges in the process.

3.2.1.3 Networking Dependencies

The network plays a crucial role in the comprehensive infrastructure solution for any organization. It is essential to recognize the reliance on the foundational networking infrastructure present in your environment. VMware's primary offering related to NSX facilitates software-defined networking (SDN), enabling the virtualization of vital network components such as switches, routers, and firewalls for infrastructure deployment. Understanding the reliance of any environment on software-defined networking requirements is vital when exploring alternatives. Important networking dependencies to consider include the following:

- **vSwitch configuration**: VMware uses virtual switches (vSwitches) to manage networking for VMs. Correctly configuring vSwitches, including VLANs and port groups, is vital for network traffic management.

- **Load balancing**: Depending on the scale of operations, organizations might need to implement load balancing across network resources to prevent bottlenecks and optimize performance.

- **Security measures**: Implementing firewall rules, intrusion detection systems, and secure access controls is essential for protecting VMware environments. Proper segmentation and isolation of networks help mitigate security risks.

Careful identification of network dependencies allows organizations to better look for different possible solutions and vendor selections.

3.2.1.4 Environmental Considerations

In addition to hardware and software dependencies, various environmental factors can influence the deployment and functioning of different virtualization solutions. The operating environment requirements are not the same for servers deployed in datacenters as compared to servers deployed at the edge locations. Apart from that, the other environmental considerations could be around the requirements for disaster recovery, etc. The key environmental considerations should include the following:

- **Power and cooling**: The requirement of power and cooling is different for solutions hosted in data centers and solutions hosted at the edge locations. When looking for possible alternative solutions, the understanding of these requirements is important as well to maintain operating conditions for physical servers.

- **Disaster recovery and backup solutions**: Solutions like VMware Site Recovery Manager (SRM) can automate recovery processes in case of failures. This is another challenge that needs to be carefully thought through when organizations should implement backup and disaster recovery strategies specifically designed for their virtual environments.

- **Any other support tools**: To maintain performance and resolve issues proactively, organizations should have monitoring and different support tools in place. VMware offers tools like SALT Stack or Aria automations, Aria Operations, Aria Operations for

Logs, etc. for management, ensuring that resources are utilized efficiently. This is another hurdle where dependencies and isolations are required when looking for different possible alternative solutions.

Recognizing the dependencies associated with VMware is essential for any organization aiming to implement virtualization successfully. These dependencies encompass hardware and software requirements, networking aspects, and environmental factors, each contributing significantly to providing an effective alternative to VMware solutions. As companies progress through their digital transformation journeys, understanding these dependencies will enable them to optimize their virtualization infrastructure, improve performance, and meet their strategic IT objectives. Through careful planning and management, organizations can fully leverage the diverse virtualization technologies at their disposal, fostering innovation and enhancing operational efficiency.

3.2.1.5 Skill Set and Training Needs

VMware has maintained a leading position in the virtualization market for an extended period, resulting in numerous IT teams developing their skills around its technologies. Shifting to a different solution may necessitate the upskilling of existing personnel or the recruitment of new employees, which could incur delays and increased expenses. Additionally, the prevalence of VMware-centric tools and processes may hinder teams' ability to adjust to an entirely new environment. Employees may also have their own preferences and reasons for remaining with a long-established solution that they are accustomed to managing. Each virtualization platform comes with its own learning curve, and moving away from VMware can expose skill gaps within your IT team. Employees trained extensively on VMware may struggle to adapt to new interfaces, processes, and underlying technologies of alternative solutions, potentially causing operational disruptions.

While this is a well-known concern and is inherent with any technology that is available from generations. But it is also important to keep your business relevant to the present industry situations. It is also important to consider investment in training and reskilling initiatives are important as well and are tailored to the alternative solution you are considering. Encouraging continuous learning equips your team with the skills needed to ensure a smoother transition and minimizes downtime during the changeover.

3.2.1.6 Data Migration Challenges

Transitioning data from VMware to a different platform presents a significant challenge characterized by various complexities. Organizations may encounter risks such as potential data loss, system downtime, or the necessity for supplementary tools to aid in the migration. It is essential to ensure data integrity and to minimize disruptions during this transition. The task of moving data from a VMware environment to an alternative platform can be intimidating and fraught with potential pitfalls. Maintaining data integrity and security throughout the migration is imperative, necessitating meticulous planning and execution. This aspect can pose an additional challenge when exploring alternative technological solutions. A thorough evaluation of various migration options, along with an understanding of the business's downtime requirements associated with each option, is crucial when seeking alternatives. It is vital to develop a comprehensive migration strategy that encompasses strong data backup, validation procedures, and a contingency plan to address any unexpected complications. Utilizing automated tools can also enhance and simplify the data migration process.

The task of exploring alternatives to VMware can prove to be quite challenging for organizations, as it involves navigating a variety of obstacles. This includes comprehending existing dependencies, overseeing data migration, and addressing financial implications.

Enterprises must undertake this process with careful consideration. Nevertheless, through meticulous planning, proactive training, and a thorough assessment of available options, organizations can achieve a successful transition. This transition can enable them to leverage a more suitable virtualization solution that aligns with their changing requirements. As technology progresses, maintaining flexibility and openness to change will be essential for organizations aiming to enhance their virtual infrastructure. It is important to recognize that each challenge offers a chance for development, and with appropriate strategies, businesses can discover effective alternatives that not only fulfill their needs but also improve their operational efficiency.

3.3 Navigating Technical Challenges and Potential Alternatives

When exploring alternatives to VMware, many organizations may find that a multi-vendor strategy is more advantageous than a single alternative, given the diverse use case scenarios. As technical challenges arise in the search for the most suitable alternative, infrastructure professionals must develop and articulate a contingency plan that aligns with their workload distribution and application strategies. It is crucial for these professionals to adopt a realistic perspective that recognizes VMware's dominant position in the server virtualization sector. Companies need to be aware of the intricate web of dependencies on VMware to make informed decisions. Figure 3-2 presents a non-exhaustive list of potential migration paths that could be taken during a large-scale migration and/or IT modernization initiative when looking for VMware alternatives.

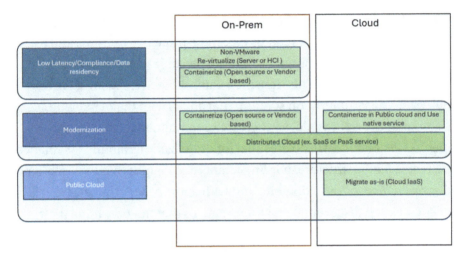

Figure 3-2. *Potential migration paths*

Let us examine the various migration paths in greater detail and conduct a brief analysis of each.

3.3.1 Use Case: Low Latency/Compliance/Data Residency Requirements

When evaluating different migration paths, there may be a situation where the requirement from the business is to keep low latency or have requirements to stay on-premises for compliance reasons. When the business requirs these workloads to stay on-premises, in those situations, apart from keeping it on "VMware," the approach could be either

- **Re-virtualize**: Virtual machine workloads can be deployed on alternative server virtualization platforms or hyper-converged infrastructure (HCI) solutions provided by various vendors. Notably, several vendors, such as Nutanix, RedHat, and Microsoft, offer hyper-converged infrastructure (HCI) solutions.

- **Containerize**: Wherever feasible, VMware consumers can look for modernizing their applications and convert them to containers. There are many container-based platforms available, both open source (Docker) and vendor organizations (RedHat OpenShift) that allow support and services for container platform technologies.

3.3.2 Use Case: Modernization

Another potential path for migration of workloads is to do the modernization. Microservices architecture and container-based technologies are increasingly getting attention worldwide from technical communities and businesses. Based on the budget and other business requirements, containerization and hosting technology can be solutioned in on-premises or in public cloud. Today almost all of the modern public cloud hosting platforms offer containers as a service where organizations can easily leverage these technologies and modernize their existing applications.

3.3.3 Use Case: Public Cloud Strategy

The growing trend of cloud adoption among numerous organizations may have been further bolstered by Broadcom's acquisition of VMware, particularly for those entities that have long contemplated a cloud strategy. One viable approach is to utilize public cloud infrastructure as a service (IaaS), platform as a service (PaaS), or software as a service (SaaS) solutions. Currently, various cloud providers, including Microsoft Azure, Google Cloud Platform (GCP), and Amazon, present a range of options that enable workloads to maintain their existing IP addresses. Additionally, these cloud platforms offer advanced migration tools, facilitating a seamless transition of workloads with minimal disruption to business operations. Organizations transitioning to the cloud are increasingly embracing native cloud solutions that align closely with their operational requirements.

The above-mentioned potential migration paths require detailed planning and encompass considerable efforts and associated risks. As there may not be a single migration path suitable for organizations, these migration paths may require large-scale migrations, which in turn require considerable planning and effort. These efforts are unique per organization based on the nature of workloads and their business demand. With regards to efforts required for different potential migration paths, here are some sample efforts and associated risks for the different choices available. Please refer to Figure 3-3 on different potential approaches mentioned in the Gartner report.

Low ○ ◔ ◑ ◕ ● High

	Migration Effort	Migration Cost	Migration Risk	Time (Duration)	Expertise (Availability of Skills)	Technical Fit	Modernization Benefit	Business Benefit
Stay with VMware (no change)	N/A	N/A	N/A	N/A	●	●	No Change	No Change
Revirtualization	◑	◑	◔	◑	◔	◕	○	○
IaaS lift-and-shift	◑	◑	○	◕	◕	◕	◑	◔
IaaS lift-and-optimize	●	◕	◔	●	◑	◑	●	◑
Cloud-hosted VMware	◔	◑	○	◔	○	●	◔	No Change
Containerization (existing workloads)	●	●	◕	●	◔	◔	◕	◑
Container-VM convergence	◕	◕	●	◕	○	◕	◑	◔
Cloud-inspired on-premises infrastructure modernization (private cloud)	◕	◕	◑	●	◑	◕	◑	◑
Cloud-hosted virtualization (non-VMware)	◕	◑	◕	◕	○	◔	○	○
Distributed cloud IaaS	●	◕	◑	◕	○	◕	◑	◑
Retire/replace (PaaS, SaaS)	●	◕	◑	◕	◑	○	●	◕

Figure 3-3. Potential migration path and associated risks/challenges

Now that we have understood the different possible migration paths, let us now take a close look at some of these challenges and possible solutions to navigate these technical challenges.

3.3.4 Data Migration Complexity

One of the most significant technical challenges when exiting VMware is migrating data from a VMware infrastructure to a new platform. VMware environments utilize formats and protocols that may not be directly compatible with other virtualization technologies or cloud services. Data migration might involve extensive planning, especially for large volumes of data that require not only transfer but also transformation. The suggested solution to mitigate the migration complexity could be

- **Assess compatibility:** Conduct a thorough analysis of both the source (VMware) and target environments to identify potential compatibility issues early in the migration process.

- **Utilize migration tools:** Leverage specialized migration tools designed to facilitate transitions from VMware to alternative platforms, which can automate many aspects of the migration process.

3.3.5 Application Dependency and Reconfiguration

Many organizations run applications that rely on the specific ecosystem provided by VMware, including VMware tools and drivers. Exiting VMware might necessitate reconfiguring applications to operate in a different environment, which can be complex and time-consuming. To help with this, the solution could be

- **Inventory and analyze applications:** Cultivate an inventory of all applications operating within your VMware environment. Understand their dependencies and configurations to devise a tailored strategy for reconfiguration.

- **Staging environment:** Create a staging environment to test applications after migration but before fully deploying them in the new environment. This allows for troubleshooting without disrupting business operations.

3.3.6 Networking and Security Configurations

VMware environments often have custom network configurations, including virtual switches, firewalls, and security policies. Transitioning to a new environment can disrupt established networking and security setups, leading to security vulnerabilities and performance issues. The possible solutions are

- **Analyze networking configurations:** Document existing network and security setups to facilitate a smooth transition to new architecture.

- **Implement security best practices:** During the migration, prioritize security by ensuring that equivalent firewall rules, policies, and access controls are in place in the new environment.

3.3.7 Legacy Infrastructure Compatibility

Organizations may have legacy systems and applications that are tightly coupled to VMware technologies. Transitioning these older systems to a new environment often requires additional resources and time. The possible solutions are

- **Evaluate legacy systems:** Identify legacy systems that are critical to operations and develop a phased migration plan.

- **Consider modernization options:** When possible, explore opportunities to update or modernize legacy applications rather than merely migrating them.

The different migration paths discussed here may not be a full list of options available for organizations. There may be some settings unique to each environment that are different but better suited to them. However, to provide more insights and suggested suitability of the workloads, Table 3-1 provides comparable inputs into different migration paths and potential benefits/challenges associated with the different options.

Table 3-1. *Potential migration path and associated suitability*

Technology	Approach	Advantages	Challenges	Suggested suitability
Hypervisor	Revirtualize	* Limited benefit but can save cost.	* Administrative experience and operational burden * Support for enterprisescale requirements * Migration of workloads and planning	* Server Consolidation and optimization
HCI	Rehost and/or Revirtualize	* Integrated management experience * Better productivity	* Learning new technology * Limited hardware and technology vendor * Cost - As Compute and storage invariable demand can cause addition of hosts to the solution	* Remote/branch office * Basic andmoderate server virtualization * Software-defined infrastructure
Public cloud IaaS	Rehost	*Access to increased agility and innovation *Pay as you go pricing	* Network latency * Cost management may pose challenges	* Lift and Optimize workloads
Containers	Rearchitect or rebuild	*Increased agility *Alignment with modern application architecture	* Requirement for cultural change * Engineering effort * Monolithic workloads	* New custom enterprise applications
Public cloud Paas, Saas	Replace	* Reduction in technical debt * Shift of I&O responsibilities to cloud provider	* Compliance mandates * Disruption to existing business processes * Availability of required functionality * Vendor lock-in	* Strategic business intiatives *Retirement of legacy workloads

Exiting VMware can be an intricate process filled with technical challenges. From data migration complexities to application dependencies, organizations must approach this transition with careful planning and strategy. By understanding and proactively addressing these hurdles, businesses can successfully navigate their exit from VMware and embrace new, potentially more suitable technologies that align with their current and future goals. Whether it's cost savings, improved performance, or greater flexibility, the advantages of a well-thought-out exit strategy far outweigh the challenges involved.

3.4 Navigating Challenges Around Upskilling Teams and Training for VMware Exit

As the technology sector advances at an extraordinary rate, organizations are increasingly confronted with the necessity to consistently enhance their skills and technologies. This challenge is especially significant for teams employing platforms such as VMware, particularly in the context of potential transitions or departures. Whether the decision to move away from VMware stems from a strategic realignment or changes in the market, the critical need for upskilling your workforce remains paramount. In this section, we will examine the difficulties associated with upskilling and training during a VMware exit and provide strategies to effectively address these challenges.

3.4.1 Understanding the Need for Upskilling

3.4.1.1 The Rapid Rise of New Technologies

The emergence of cloud computing, containerization, and serverless architectures necessitates that your team remains proactive in adapting to these advancements. VMware, historically recognized as a pioneer in virtualization, now encounters competition from firms that provide innovative solutions potentially more aligned with your organization's present requirements. Each vendor presents a distinct offering that distinguishes them in the market and may surpass the capabilities of their rivals. It is crucial to maintain a forward-thinking approach, and your organization must stay informed about the latest technologies to enhance business operations effectively.

3.4.1.2 The Implications of a VMware Exit

Exiting VMware might be driven by the need for more agile and efficient technologies or cost-saving measures. However, it also raises questions about your team's capabilities. Have they had the training to work with the new technologies or platforms you're switching to? Are they comfortable with this change, or will a knowledge gap impede your operational effectiveness?

3.4.2 Recognizing the Challenges

3.4.2.1 Resistance to Change

Change is often met with resistance, and this is particularly true in tech environments where team members have invested time to master VMware's intricacies. Some may feel their skills are no longer relevant or fear they will struggle with a new environment.

3.4.2.2 Resource Constraints

Training and upskilling require time and financial resources. Depending on your organization's budget and the urgency of the transition, this can pose a significant challenge. There's often pressure to deliver results quickly, making it difficult to dedicate enough time for comprehensive training to the alternative technology.

3.4.2.3 Diverse Skill Levels

Your team will likely possess varying levels of familiarity with new technologies. This disparity can lead to frustration and hinder cohesion. Tailoring training to accommodate different learning paces and background knowledge becomes a challenge.

3.4.2.4 Keeping Up with Continuous Changes

The tech world is not static; new tools and platforms emerge regularly. The learning process must be ongoing, which can be daunting for teams already grappling with daily tasks and project deadlines.

3.4.3 Strategies for Upskilling and Training

3.4.3.1 Develop a Clear Training Roadmap

Start by assessing the current skill levels of your team and identify the technologies and areas of expertise needed for a successful transition. With this assessment, you can build a tailored training roadmap that specifies what needs to be learned and when. Include milestones and timelines for tracking progress.

3.4.3.2 Invest in Formal Training Programs

Consider partnering with training organizations or technology vendors that provide structured learning paths. Certification courses can help validate skills and ensure your teams are up to speed with the latest technologies. Variety in training formats—such as online courses, workshops, and hands-on labs—will cater to different learning styles.

3.4.3.3 Foster a Culture of Continuous Learning

Encourage a mindset of lifelong learning within your organization. Create opportunities for knowledge sharing, whether through regular lunch-and-learns, hackathons, or team discussions. Offering incentives for completing training modules or certifications can also motivate team members to engage.

3.4.3.4 Leverage Internal Talent

Identify internal champions who have a strong understanding of the transitioning technology. Empower them to lead training sessions, which can help build confidence within the team and establish a collaborative learning environment.

3.4.3.5 Implement Mentorship Programs

Pair less experienced staff with those who have more advanced skills in the new technologies. This peer-to-peer mentoring can foster deeper understanding and provide support as employees learn to navigate new tools and methodologies.

3.4.3.6 Allocate Time for Learning

Where feasible, set aside dedicated time for training within work schedules to ensure that team members can focus on upskilling without the pressure of daily responsibilities. Consider establishing 'training weeks' during less stressful periods where teams can immerse themselves in learning without distractions.

Enhancing the skills of your team while transitioning from VMware presents a mix of challenges and opportunities. It is essential to approach this process with a well-defined strategy that recognizes the importance of ongoing education, allocates resources for formal training, and fosters a culture that prioritizes growth and adaptability. As your teams become more proficient in new technologies, your organization will be better equipped to manage the uncertainties inherent in a rapidly evolving technological landscape. View this transition not merely as a necessity but as a chance to innovate and excel.

Remember, investing in the development of your team is fundamentally an investment in the future resilience and success of your organization.

3.5 Vendor Landscape in Server Virtualization

In the dynamic realm of information technology, server virtualization emerges as a crucial technology that supports contemporary data centers and cloud infrastructures. As the demand for more flexible, cost-effective, and diverse virtualization options grows, a variety of alternatives to VMware have emerged, each offering unique features and benefits. In this section, we will explore the current vendor landscape in server virtualization, focusing on some key alternatives to VMware that organizations may consider based on their specific needs and use cases.

Server virtualization allows multiple virtual machines (VMs) to run on a single physical server, enabling better resource utilization, reduced IT costs, and improved operational efficiency. With the advent of cloud computing, server virtualization has become even more critical, facilitating seamless scalability and agile deployment of applications.

3.5.1 The VMware Dominance

VMware has positioned itself as a prominent leader in the realms of server virtualization and private cloud services, establishing benchmarks that have fundamentally altered organizational approaches to IT infrastructure. Its extensive history, broad adoption, and a strong ecosystem of third-party tools and integrations have made VMware a preferred option for enterprises globally. However, following its merger with Broadcom, VMware's prominence in the market has begun to wane. This merger has sparked concerns among customers and industry analysts regarding a potential shift in focus from VMware's core strengths to Broadcom's hardware-oriented business model. Customers fear that the focus may shift away from software innovation and customer support in favor of immediate financial returns, which could lead to a perceived decline in the quality and reliability of VMware's products. Furthermore, as Broadcom aims to incorporate VMware's operations into its larger portfolio, uncertainty surrounding product roadmaps has arisen, prompting many enterprises to reassess their dependence on VMware technologies. This strategic transition has created opportunities for rival providers to take advantage of VMware's diminishing influence, particularly in the private cloud arena, where alternatives such as Microsoft Azure Stack, RedHat OpenShift, and various native cloud solutions are gaining momentum. As organizations seek more adaptable and cost-efficient virtualization options, VMware's previously unassailable momentum seems to be slowing, raising critical questions about its future in an increasingly competitive environment that necessitates ongoing innovation and agility. The ramifications of this transformation are profound, as businesses that once heavily relied on VMware now find themselves at a crossroads, weighing the advantages of migrating to other platforms against the familiarity and reliability of their current systems. The convergence of these factors marks a crucial juncture for VMware, where sustaining its legacy as a leader is at stake.

3.5.2 Gartner's Insights on Server Virtualization

When we are discussing server virtualization, it is important to understand Gartner's insights on the server virtualization market post-Broadcom acquisition of VMware. Gartner, a leading research and advisory firm, frequently analyzes the server virtualization market and provides insights that guide organizations in their decision-making processes. Here are some key takeaways from Gartner's assessments on server virtualization market trends in the year 2024. These insights are providing indications of where the industry trend is and what to look for when evaluating the server virtualization market for an eligible alternative to the VMware portfolio.

3.5.2.1 Cloud vs. On-Premises Strategies

Although on-premises server virtualization continues to be essential for numerous organizations, Gartner observes an increasing shift toward cloud-based virtualization solutions. The prevalence of hybrid IT environments is rising, with workloads being allocated between on-premises and public cloud resources. Gartner highlights the necessity of formulating a balanced strategy that corresponds with business objectives and operational requirements.

3.5.2.2 Emergence of Hyper-converged Infrastructure (HCI)

Gartner has highlighted the growing trend of hyper-converged infrastructure (HCI), which combines computing, storage, and networking into a unified system. HCI streamlines both management and deployment processes, allowing organizations to implement and expand their virtualization technologies with greater efficiency. This trend signifies a movement toward more integrated systems capable of accommodating diverse demands and workloads.

3.5.2.3 Security Considerations

In an era characterized by swift digital evolution, the importance of security cannot be overstated. Research conducted by Gartner underscores the urgent requirement for improved security strategies in virtualized settings. It is essential for organizations to focus on establishing strong security protocols to safeguard their virtual machines and data against possible threats. This approach should encompass the use of micro-segmentation, zero-trust frameworks, and sophisticated monitoring solutions.

3.5.2.4 Cost Management and Optimization

Although virtualization has the potential to generate cost savings, Gartner recommends that organizations perform comprehensive cost-benefit analyses. It is essential to grasp the total cost of ownership (TCO) and return on investment (ROI) to ensure that virtualization technologies align with financial goals. Additionally, organizations should remain updated on licensing models and cloud pricing frameworks to enhance their expenditure management.

3.5.2.5 AI and Automation

Gartner has recognized the expanding significance of artificial intelligence (AI) and automation in the realm of server virtualization. Utilizing AI-generated insights can facilitate the optimization of resource distribution, enhance performance monitoring, and improve self-healing functionalities in virtualized settings. Furthermore, automation tools can simplify the processes of provisioning, management, and scaling, thereby minimizing manual effort and boosting overall efficiency.

3.5.3 The Vendor Landscape in Server Virtualization and Private Cloud

The private cloud sector presents a varied vendor landscape, featuring a multitude of solutions and offerings. Prominent companies such as Microsoft, Nutanix, RedHat, and numerous open-source initiatives play a significant role in shaping the competitive environment. Microsoft Azure Stack is particularly notable for organizations aiming to extend their Azure public cloud services into a private cloud setting, as it provides seamless integration and familiar development tools. This capability enables businesses to utilize their existing expertise while ensuring a consistent platform for application development and deployment. Conversely, Nutanix has established itself as a frontrunner in hyper-converged infrastructure (HCI), allowing enterprises to streamline their data center operations through a software-centric approach that merges storage, computing, and virtualization into a unified platform. Its cloud solution prioritizes user-friendliness, scalability, and adaptability, facilitating the effective adoption of a hybrid cloud strategy by organizations. RedHat plays a crucial role in this ecosystem with its OpenShift and RedHat Virtualization offerings, which focus on enterprise-level Kubernetes container orchestration and traditional virtualization, respectively. RedHat's dedication to open-source principles not only drives innovation but also promotes community collaboration, making it an appealing option for organizations that value customization and control over their infrastructure. Additionally, the open-source community enhances the vendor landscape, with projects like OpenStack offering a versatile framework for constructing and managing private clouds, enabling organizations to customize solutions to their specific requirements without being tied to a single vendor. This increasing focus on open-source solutions empowers businesses to take advantage of community-driven innovations while ensuring cost efficiency. Together, these entities foster a vibrant and competitive atmosphere that provides organizations with a variety of options to fulfill their private cloud requirements.

Figure 3-4 presents a comprehensive overview of the current vendor landscape within the industry, along with an analysis of their respective offerings related to Full Stack infrastructure solutions. We will explore these aspects in greater detail in the next section.

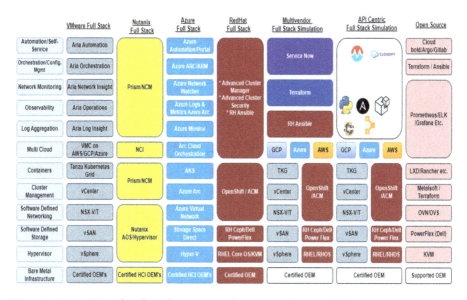

Figure 3-4. *Vendor landscape and comparison with VMware Full Stack*

3.5.3.1 Microsoft Azure Stack HCI

In recent years, organizations have increasingly adopted hybrid cloud solutions, enabling them to leverage the scalability and flexibility of cloud services while retaining control over their on-premises infrastructure. Microsoft Azure Stack HCI (hyper-converged infrastructure) has become a prominent contender in this domain. The subsequent chapter will provide an in-depth discussion of Microsoft's solution. In this section, we will outline the complete components of Microsoft Azure Stack HCI, including its latest release, 23 H2, and provide a brief overview of these elements, allowing readers to gain a clearer understanding of its capabilities and how it can enhance their hybrid cloud strategies.

What Is Azure Stack HCI?

Azure Stack HCI is a service offered by Azure that extends the flexibility of cloud computing to local environments. It integrates virtualization, software-defined storage, and networking into a cohesive solution, enabling organizations to operate their applications in a more efficient and streamlined way. The H2 2023 release from Microsoft has brought forth improvements aimed at enhancing performance, security, and overall manageability.

Azure Stack HCI OS

Central to Azure Stack HCI is its operating system, which is based on Windows Server 2022. This operating system is specifically designed for hyper-converged infrastructure, offering the fundamental elements necessary for the deployment of virtual machines (VMs), storage management, and network operations. The latest version of the OS features enhanced resilience and supports nested virtualization, thereby optimizing resource utilization and operational efficiency. Azure Stack HCI leverages Hyper-V, Microsoft's robust virtualization platform, to allow organizations to run multiple VMs on the same hardware. The 23 H2 update introduces enhanced VM management capabilities and support for Linux containers, making it easier to run a diverse range of applications.

Software-Defined Storage (SDS)

Microsoft's SDS solution is called storage space direct (SSD), offering enhanced storage virtualization capabilities. The 23 H2 update provides new features such as support for storage tiers, improved deduplication, and compression. This combination allows organizations to optimize their storage needs while maximizing cost efficiency.

Software-Defined Networking (SDN)

Networking represents a vital aspect of Azure Stack HCI, which incorporates sophisticated software-defined networking functionalities. The H2 release bolsters network security by introducing features such as Azure Firewall integration and enhanced network segmentation, thereby creating a more resilient and secure environment for applications.

Azure Arc Integration and Windows Admin Center Support

Azure Arc enhances the management capabilities of Azure by extending them to on-premises settings and edge locations. The 23 H2 version enables users to effortlessly integrate Azure Stack HCI with Azure services, facilitating cohesive management, governance, and security across hybrid environments. Organizations are empowered to utilize Azure policies for direct resource management from the Azure portal.

WAC provides a central place to manage Azure Stack HCI resources easily. The latest version includes improved UI capabilities for managing clusters and VMs, along with enhanced monitoring and reporting features. These updates ensure that IT administrators can keep an eye on their workloads and resources with ease.

Azure Storage Systems

Integrating with Azure Storage services enhances the capabilities of Azure Stack HCI. This allows organizations to leverage Azure Blob Storage, File Storage, and other services for backup, archiving, and disaster recovery. The new version streamlines this integration to facilitate easy data migration and management.

Security

Security is a top priority for organizations leveraging hybrid cloud solutions. The H2 2023 update brings several security enhancements, including Secure Core Server features, which help protect against firmware vulnerabilities. Moreover, the latest version includes built-in encryption options for data at rest and in transit, making compliance with regulatory standards easier.

Monitoring and Troubleshooting Tools

Azure Stack HCI 23 H2 comes with enhanced monitoring tools that provide deeper insights into system performance and troubleshooting. These tools utilize Azure Monitor and Log Analytics to give administrators visibility into the health of their infrastructure, allowing them to proactively address issues.

Licensing Flexibility

Microsoft has recognized that organizations vary in their licensing needs, so the HCI platform allows the user of the hybrid benefit, where customers can exchange their Windows Server Datacenter licenses with HCI deployment, which results in a $0 monthly charge for the solution. In addition to that, Hybrid benefit also allows customers to run the guest windows VM's for free. This is overall a tremendous benefit that can result in reducing the overall cost of the entire solution and better ROI.

With its extensive array of features and components, Microsoft Azure Stack HCI 23 H2 stands out as a powerful solution for organizations aiming to leverage the advantages of hybrid cloud environments. Its strong virtualization capabilities and seamless integration with Azure are designed to optimize operations and enhance efficiency. By comprehensively understanding the offerings in this latest version, IT professionals can more effectively plan their infrastructure investments to meet the increasing demands for agile, secure, and scalable IT solutions.

3.5.3.2 Nutanix Full Stack Components

Organizations are persistently in pursuit of strategies that enable them to enhance operational efficiency, minimize expenses, and optimize productivity. A notable integrated solution that has gained recognition is Nutanix, a frontrunner in hyper-converged infrastructure (HCI). Nutanix facilitates the management of data, ensures data protection, and streamlines application usage, allowing businesses to fully harness the capabilities of their IT ecosystems. This discussion will delve into the fundamental elements of the Nutanix Full Stack, outlining their functions and advantages.

What Is Nutanix?

Nutanix is a cloud software enterprise that provides a hyper-converged infrastructure platform aimed at streamlining and enhancing data management processes. This platform consolidates computing, storage, networking, and virtual machine management into a unified solution, enabling organizations to efficiently oversee their workloads across public, private, and hybrid cloud environments. The comprehensive stack of Nutanix consists of various interconnected elements, each tailored to offer distinct functionalities and generate value throughout the organization. Let us explore each of these components in brief below.

Nutanix Acropolis (AOS)

At the heart of Nutanix's architecture is the Acropolis Operating System (AOS). AOS provides the foundational software layer, managing compute, storage, and networking resources. It offers key features such as

- **Distributed storage fabric**: AOS allows for optimal data usage by leveraging a scale-out storage architecture across nodes.

- **Data protection**: It provides native replication and backup solutions, ensuring data durability and availability.

- **Self-healing**: The system can automatically detect and recover from failures, minimizing downtime and enhancing reliability.

Nutanix Prism

Nutanix Prism is the management interface that gives users visibility and control over their Nutanix environment. Available in both a web and mobile application, Prism simplifies the administration of the infrastructure with features such as

- **Single pane of glass management**: Admins can manage all components of their IT environment from a unified dashboard.

- **Intelligent operations**: Prism employs machine learning to offer predictive insights, enabling proactive resource management.

- **Performance monitoring**: Users can monitor performance metrics in real-time, helping to identify bottlenecks and optimize resource allocation.

Nutanix Calm

Calm is Nutanix's application lifecycle management solution. It allows organizations to automate the deployment, management, and scaling of applications across any environment. Key features of Calm include:

- **Blueprints**: Users can define and automate application deployment with reusable blueprints, reducing the time to provision resources.

- **Multi-cloud management**: Calm supports deployments across multiple clouds, making it easier for organizations to manage hybrid environments.

- **Self-service provisioning**: It empowers users to provision resources on demand while adhering to governance policies.

Nutanix File and Object Storage

Nutanix also offers integrated file and object storage solutions that complement its HCI platform. This component provides

- **Native file services**: Organizations can store and manage unstructured data alongside structured data within the same platform.

- **S3-compatible object storage**: The solution supports applications that require object storage, extending Nutanix's capabilities into modern data architectures.

Nutanix Flow

Security and data protection are critical in today's digital landscape, and Nutanix Flow addresses these needs by providing advanced security features. Key functionalities include

- **Micro-segmentation**: Flow allows organizations to isolate and secure workloads by applying granular security policies.

- **Visibility and analytics**: Flow provides insights into traffic patterns and potential security threats, enabling proactive management.

Nutanix Cloud Native Services

As organizations increasingly adopt cloud-native technologies, Nutanix supports this transition with integrated cloud-native services. Features include

- **Kubernetes support**: Nutanix offers built-in Kubernetes management, allowing organizations to deploy, manage, and scale containerized applications seamlessly.

- **Integration with CI/CD pipelines**: The platform integrates with popular CI/CD tools, facilitating agile development practices.

Nutanix Cloud Clusters (NC2)

For organizations eager to leverage public cloud resources without compromising on control and ease of management, Nutanix Cloud Clusters (NC2) provides a seamless extension of Nutanix environments to the public cloud. Key benefits include

- **Familiar management experience**: Clients maintain the same management interface and operational model when working across on-premises and public cloud infrastructures.

- **Bursting capabilities**: Workloads can be easily scaled up or down based on demand, optimizing resource utilization and reducing costs.

The Nutanix Full Stack is crafted to equip organizations with a holistic strategy for managing their IT infrastructure. By integrating compute, storage, networking, and application management seamlessly, it allows businesses to enhance performance, scalability, and resilience in their operations. As organizations increasingly pursue digital transformation, comprehending and utilizing the elements of the Nutanix platform can significantly contribute to fostering efficiency and innovation. Whether the goal is to modernize a data center or implement a hybrid cloud strategy, Nutanix offers the necessary tools to fulfill these objectives. In today's ever-evolving landscape, establishing a strong and adaptable infrastructure is crucial, and Nutanix presents an opportunity to streamline complexity and realize the full capabilities of your IT environment.

3.5.3.3 RedHat Full Stack

RedHat, a prominent provider of open-source solutions, presents a robust Full Stack Private Cloud solution tailored to address the varied requirements of enterprises. This discussion will examine its essential components and how they collaboratively contribute to a holistic private cloud experience. The RedHat Full Stack Private Cloud serves as an all-encompassing solution that includes all the vital elements necessary for constructing, managing, and sustaining a private cloud environment. Its objective is to offer the agility characteristic of public clouds while maintaining enterprise-level governance over data and infrastructure. In addition to its open-source foundation, it emphasizes the integration of diverse technologies to create a unified, adaptable, and scalable private cloud.

RedHat OpenShift

Central to the Full Stack Private Cloud solution is RedHat OpenShift, an enterprise-grade Kubernetes platform designed to streamline the management of containerized applications. OpenShift aids developers by facilitating continuous integration and continuous deployment (CI/CD) methodologies, thereby promoting swift application development. It accommodates a hybrid and multi-cloud environment, enabling organizations to deploy applications effortlessly across diverse platforms.

RedHat OpenStack Platform

The RedHat OpenStack Platform serves as a powerful infrastructure-as-a-service (IaaS) solution, enabling organizations to effectively manage their compute, storage, and networking resources. OpenStack is particularly well-suited for the development of scalable and adaptable cloud environments, providing features such as virtual machines, block storage, and networking-as-a-service. This platform empowers businesses to deploy services and applications in a manner that is both controlled and scalable.

RedHat Ansible Automation Platform

Automation plays a vital role in contemporary IT operations. The RedHat Ansible Automation Platform empowers organizations to streamline their cloud management and operational processes, thereby increasing productivity. Its user-friendly interface and robust orchestration features enable teams to deploy and oversee IT services with minimal manual effort. This leads to more efficient deployments, faster incident resolution, and enhanced overall effectiveness.

RedHat Ceph Storage

Reliable and scalable storage is a fundamental requirement for any cloud infrastructure. RedHat Ceph Storage offers a highly adaptable and scalable storage solution capable of managing diverse workloads. It can be implemented as object storage, block storage, or file storage, all while ensuring redundancy and high availability. Furthermore, Ceph integrates effortlessly with OpenStack, enabling organizations to utilize a unified storage backend for all their cloud requirements.

RedHat Virtualization

This component enables organizations to establish virtualized environments for the operation and administration of virtual machines. RedHat Virtualization is founded on the open-source KVM (kernel-based virtual machine) technology, offering a dependable platform for both conventional workloads and contemporary cloud-native applications. It empowers IT teams to optimize resource utilization while streamlining workload management.

RedHat Insights

Achieving operational excellence within a private cloud necessitates ongoing monitoring and the use of predictive analytics. RedHat Insights serves as a proactive management solution that delivers automated oversight and reporting regarding the condition of your cloud infrastructure. By utilizing data and analytics, Insights enables IT teams to detect potential problems prior to their escalation, thereby maintaining a stable and high-performing environment.

RedHat OpenShift Service Mesh

The management of microservices and their interconnections presents various challenges. The OpenShift Service Mesh, which utilizes Istio, offers sophisticated traffic management, security, and observability features for microservices. This enables organizations to develop robust applications with regulated access and enhanced security while also obtaining critical insights into the functioning of their services.

RedHat Advanced Cluster Management for Kubernetes

As organizations oversee numerous Kubernetes clusters in various environments, the necessity for centralized management becomes evident. The Advanced Cluster Management tool enhances the administration and visibility of multiple clusters, regardless of whether they are hosted on-premises or in the cloud. It empowers teams to effectively manage resources, applications, and security policies across a range of environments.

The RedHat Full Stack Private Cloud solution combines technologies like OpenShift, OpenStack, and Ansible with advanced storage and monitoring to create a flexible, scalable, and fully managed private cloud infrastructure. By leveraging RedHat's open-source approach, organizations can customize and expand their cloud solutions to meet evolving needs. This solution enhances agility, cost-effectiveness, and security, positioning enterprises for future success. Investing in a RedHat Full Stack Private Cloud can be a crucial step for organizations looking to take control of their cloud strategies.

3.5.3.4 Open-Source Solutions

Developing a comprehensive open-source full-stack infrastructure solution for private cloud environments requires a meticulous integration of diverse components that promote scalability, flexibility, security, and effective resource management. Central to this infrastructure is the virtualization layer, where hypervisors such as KVM (kernel-based virtual machine) and Xen provide essential abstraction of physical hardware, allowing multiple isolated instances to operate on a single machine. These hypervisors should be supported by centralized management tools like OpenStack or Apache CloudStack, which deliver intuitive dashboards for resource provisioning, usage monitoring, and automation of virtual machine and service deployment. Additionally, containerization technologies such as Docker, paired with orchestration frameworks like Kubernetes, significantly enhance the architecture by facilitating microservices designs that can scale applications horizontally, thereby simplifying updates and maintaining system integrity. For storage, integrating solutions like Ceph or GlusterFS offers scalable, distributed storage capabilities that ensure data redundancy and high availability, crucial for private cloud environments. Networking components, including Open vSwitch and Calico, provide flexible and secure networking options that improve traffic management and isolation among various user environments. In terms of security, implementing tools like HashiCorp Vault for secrets management and Encrypt for automated SSL/TLS certificates strengthens the overall security framework of the private cloud. Lastly, monitoring and logging tools such as Prometheus for performance tracking and the ELK stack for log management are vital for maintaining the health of the infrastructure. By utilizing these open-source components, organizations can establish a thorough private cloud solution that not only addresses their specific operational requirements but also encourages innovation through the adaptability of open-source technology.

Automation/Self-service

When selecting a cloud-based automation self-service platform, three key options stand out: Cloud Bold, Argo, and GitLab. Each offers unique benefits for enhancing continuous integration and delivery (CI/CD) processes.

Cloud Bold features an intuitive interface and strong integration capabilities, allowing teams to automate workflows easily and manage their own deployments, which speeds up the development lifecycle. Argo, designed for Kubernetes, excels in containerization, providing scalable automation through tools like Argo Workflows and Argo CD, which facilitate complex workflow management and version control. GitLab integrates source code management with CI/CD, simplifying the development pipeline and enhancing team collaboration by allowing developers to create pipelines directly from the repository.

Ultimately, the choice among these platforms depends on an organization's specific needs and infrastructure. Regardless of the selection, using any of these tools can lead to improved efficiency, reduced operational costs, and greater agility in responding to market demands, strengthening competitive advantage in the digital landscape.

Orchestration/Configuration Management

Terraform and Ansible are key tools for orchestration and configuration management in modern IT, particularly in cloud and DevOps environments. Terraform, an open-source infrastructure as code (IaC) tool by HashiCorp, allows users to define and provision infrastructure using HashiCorp Configuration Language (HCL). It automates resource management across cloud providers like AWS, Azure, and Google Cloud, overseeing the entire infrastructure lifecycle while minimizing manual errors through a state file. In contrast, Ansible, developed by RedHat, focuses on configuration management and application deployment with an agentless architecture. It uses YAML-based playbooks to execute tasks across systems via SSH or API calls, ensuring consistent configurations and software installations. Together, Terraform and Ansible provide a comprehensive solution for automating infrastructure provisioning and application deployment, enhancing efficiency, speeding up deployment, and improving collaboration between development and operations teams.

Observability Stack

Prometheus, ELK (Elasticsearch, Logstash, Kibana), and Grafana are three powerful tools widely utilized in modern observability stacks, each serving a distinct yet complementary purpose in monitoring, logging, and visualization for IT infrastructure and applications. Prometheus excels in monitoring and alerting, designed to track time-series data through a pull-based model that scrapes metrics from instrumented jobs at specified intervals. Its powerful query language, PromQL, allows users to aggregate and analyze metrics in real time, making it an ideal choice for performance monitoring and system health checks. On the other hand, the ELK stack specializes in log management and analysis. Elasticsearch serves as the robust search and analytics engine, Logstash processes and ingests log data from various sources, and Kibana provides a web interface for visualizing and querying the data stored in Elasticsearch. This triad enables organizations to capture vast amounts of log data, perform powerful searches, and visualize trends and anomalies, thereby providing deep insight into application performance and user behavior. Complementing these tools, Grafana acts as a leading visualization platform that can integrate seamlessly with both Prometheus and Elasticsearch, allowing users to create dynamic dashboards that bring together different data sources. With Grafana, teams can visualize metrics from Prometheus alongside logs from the ELK stack, facilitating an efficient and comprehensive approach to observability. The synergy between Prometheus, ELK, and Grafana empowers organizations to enhance their monitoring capabilities and develop a robust observability framework, ultimately leading to improved system reliability and faster incident response times, which are crucial in today's fast-paced digital landscape. Through the integration of these technologies, development and operations teams can more effectively gain insights into their systems, identify bottlenecks, and drive performance optimization initiatives while also ensuring compliance and security by monitoring logs and system behavior closely.

Container Management and Orchestration

LXD and Rancher are powerful tools that can significantly enhance the management and orchestration of containers and virtual machines in modern IT environments. LXD, a container hypervisor developed by Canonical, is designed to provide a more advanced set of features for managing LXC containers, allowing for a higher level of abstraction and control than traditional container solutions. It brings the ability to launch and manage lightweight, resource-efficient Linux containers that feel like virtual machines, complete with their own filesystem, network, and processes, thus enabling developers and system administrators to run applications in isolated environments without the overhead typically associated with full virtual machines. On the other hand, Rancher serves as a comprehensive container management platform that simplifies the deployment, management, and scaling of Kubernetes clusters, offering integrated tools for monitoring, networking, and security. Rancher allows teams to manage multiple Kubernetes clusters across various environments, from on-premises data centers to cloud services, all from a single interface. The LXD and Rancher can create a robust infrastructure for deploying microservices, facilitating development workflows, and optimizing resource utilization. By leveraging LXD's lightweight containers within the Rancher ecosystem, organizations can achieve enhanced performance and quicker deployment times while maintaining the flexibility to scale their applications as demand increases. Overall, LXD and Rancher illustrate a forward-thinking approach to container management that empowers developers to build scalable, resilient applications while simplifying the operational complexity that often comes with traditional virtualization and orchestration solutions.

Cluster Management

Metalsoft represents a cutting-edge solution for managing clusters, effectively tackling the complexities and challenges inherent in the deployment and maintenance of large-scale distributed systems. As organizations increasingly depend on cluster computing for diverse applications—ranging from data analytics to machine learning and high-performance computing—the efficient management of these clusters becomes paramount. Metalsoft meets this need by offering a comprehensive orchestration layer that streamlines the provisioning, scaling, and monitoring of clusters across various environments. Its intuitive interface enables system administrators to configure resources effortlessly, automate workflows, and enhance performance, while its robust backend integrates flawlessly with widely used container orchestration platforms such as Kubernetes. This integration facilitates dynamic resource allocation, allowing organizations to swiftly adapt to fluctuating workloads and requirements. Additionally, Metalsoft features advanced capabilities such as intelligent load balancing, fault tolerance, and real-time analytics, which not only improve operational efficiency but also ensure that clusters remain resilient and responsive. Security is a significant focus, as Metalsoft equips users with extensive tools for managing access controls and threat detection, enabling teams to protect sensitive data while minimizing potential vulnerabilities. In summary, Metalsoft transforms cluster management by merging user-friendliness with powerful functionalities, ultimately empowering enterprises to maximize their computational resources while lowering overhead and operational expenses.

Software-Defined Networking

Open vSwitch (OVS) and Open Virtual Network (OVN) are key components of software-defined networking (SDN), revolutionizing network management and orchestration. OVS acts as a multilayer virtual switch, enabling network automation and supporting essential protocols like VLAN tagging and tunneling for complex topologies. OVN enhances OVS by adding a logical layer for network virtualization in cloud environments, allowing dynamic provisioning of virtual networks on existing infrastructure. This separation from physical hardware increases flexibility in application deployment and enables reconfiguration without hardware changes. OVN also provides a northbound API for developers to manage network policies and connectivity, fostering automation and orchestration. As organizations adopt cloud-native architectures, OVS and OVN are crucial for aligning network management with the demands of modern applications. Their support for overlays, security groups, and distributed routing helps optimize performance and enhance security in multi-tenant environments, making their collaboration essential for building scalable and efficient networks.

Software-Defined Storage

PowerFlex, a product of Dell Technologies, signifies a significant advancement in software-defined storage (SDS), catering to the dynamic requirements of contemporary data environments with remarkable agility and efficiency. As organizations face the rapid expansion of data and the increasing necessity for adaptable infrastructure, PowerFlex presents a formidable solution that separates storage from hardware, allowing businesses to scale their storage capabilities independently from their computing resources. This feature not only improves the overall efficiency and utilization of IT assets but also promotes a more efficient method of managing workloads across diverse environments, whether on-premises, in the cloud, or in hybrid setups. With its sophisticated architecture, PowerFlex employs a distributed design that ensures high performance and resilience, guaranteeing that data remains accessible and secure regardless of the underlying infrastructure. The platform boasts a variety of industry-leading functionalities, including dynamic provisioning, automated data placement, and intelligent management features that further enhance storage operations. Additionally, PowerFlex integrates effortlessly with both existing systems and popular orchestration tools, offering users the flexibility to implement a hybrid cloud strategy without sacrificing performance. As organizations continue to innovate and embrace digital transformation initiatives, PowerFlex emerges as a vital enabler of agility, equipping businesses to swiftly adapt to evolving demands while retaining control over their storage resources. By utilizing PowerFlex, companies can not only improve their operational efficiency but also position themselves to seize new opportunities in an increasingly data-driven landscape, making it an indispensable element of any progressive IT strategy.

Server Virtualization

KVM, which stands for kernel-based virtual machine, has emerged as a powerful and versatile open-source server virtualization technology that fundamentally transforms the way organizations deploy and manage their IT resources. Built into the Linux kernel, KVM leverages the performance and capabilities of the underlying operating system, transforming it into a hypervisor that enables the creation and management of multiple virtual machines (VMs) on a single physical server. This capability not only optimizes resource utilization but also enhances scalability and flexibility, allowing businesses to quickly adapt to changing demands. One of the hallmark features of KVM is its compatibility with a wide range of Linux distributions and its support for various operating systems as guest VMs, including Windows and other Unix-like systems. Furthermore, KVM provides robust security and isolation for each VM, making it a favorite choice for cloud computing environments, data centers, and enterprises looking to consolidate their workloads while maintaining high levels of performance and reliability. The architecture of KVM, which separates management and virtualization functions, allows for excellent integration with orchestration and management tools, such as OpenStack, leading to streamlined deployment and configuration processes. Additionally, KVM benefits from an active community of developers and users who continuously contribute to its enhancement, ensuring that it remains up to date with the latest technological advancements in virtualization and cloud computing. The fact that KVM is open source means that organizations can access and modify the source code according to their specific needs, fostering innovation and customization that proprietary solutions may not provide. As businesses continue to embrace virtualization to reduce costs and increase operational efficiency, KVM stands out as a leading option, embodying the principles of performance, flexibility, and cost-effectiveness that are essential in today's dynamic IT landscape. With its robust feature set, active community support, and seamless integration capabilities, KVM is an excellent choice for organizations looking to harness the full potential of server virtualization while maintaining the freedom and control that comes with open-source software.

3.6 Summary

In this chapter, we emphasized the implications of Broadcom's resistance, which reflects a desire for increased revenue while showing limited regard for market dynamics and consumer requirements. In our exploration of alternatives to VMware, we also examined various factors that warrant careful consideration, which are essential for developing an effective strategy for transitioning away from VMware. Additionally, this chapter proposed several potential use cases and options for large-scale migration while also addressing the various efforts required for each alternative. We recognized that post-VMware, customers may face a landscape devoid of a singular vendor technology, necessitating a blend of diverse modern technologies. Lastly, the chapter offered insights into the current vendor landscape within the server virtualization sector, aiming to deliver a thorough comparison of various technologies and vendor offerings against the VMware product suite. The objective of this chapter is to equip readers with the necessary information to navigate the challenges arising from Broadcom's acquisition of VMware, ensuring their journey is informed and enriched with relevant insights.

CHAPTER 4

Managing VMware Dependence

In the contemporary digital economy, characterized by rapid change, businesses must prioritize agility and innovation to maintain a competitive edge. However, many organizations inadvertently find themselves reliant on a single technology provider, often overlooking the associated risks. VMware has long been a fundamental component of IT infrastructure, underpinning datacenters, cloud strategies, and enterprise workloads with its comprehensive virtualization solutions. Despite its numerous benefits, excessive dependence on VMware or any singular vendor can expose organizations to challenges that hinder future growth and adaptability. The dangers of vendor lock-in are significant, ranging from escalating costs to a potential decline in innovation. As IT environments transition toward multi-cloud and hybrid models, organizations must critically evaluate their reliance on VMware. Key questions arise: Are we overly dependent on VMware? Can we maintain long-term flexibility and security while ensuring operational efficiency?

The preceding chapter primarily concentrates on offering a comprehensive overview of the market dynamics within the server virtualization sector while also addressing the prevalent migration strategies in use today. This chapter serves as a strategic pointer for organization decision-makers seeking to reevaluate their VMware dependence. It explores not just the technological but also the operational

© Sumit Bhatia and Chetan Gabhane 2025
S. Bhatia and C. Gabhane, *Navigating VMware Turmoil in the Broadcom Era*,
https://doi.org/10.1007/979-8-8688-1264-4_4

and financial implications of relying heavily on a single vendor. We will lay out actionable steps for assessing alternative solutions, creating an efficient and secure migration strategy, and designing a resilient multi-vendor ecosystem. Through careful planning and execution, organizations can mitigate the risks associated with VMware dependence, enhance their scalability, and safeguard their infrastructure for the future – no matter how the IT landscape shifts.

The path to reducing dependence on VMware is not without its complexities, but with the right insights and a solid roadmap, businesses can unlock the potential of a more flexible, cost-effective, and secure IT architecture. This chapter provides that roadmap.

4.1 Technical Evaluation Criteria for Vendors and Migration Paths

In Chapter 3, we examined the various evaluation dimensions relevant to organizations exploring alternatives to VMware technology. This section focuses on the technical evaluation criteria essential for organizations aiming to reduce their reliance on VMware. Conducting a strategic and comprehensive technical assessment of alternative vendors is not merely advisable; it is an essential requirement. This evaluation serves as the cornerstone for a successful transition, ensuring that the organization selects a platform capable of addressing its current requirements while also offering the flexibility and scalability necessary for future expansion. The process necessitates meticulous consideration of numerous factors, each of which significantly influences the organization's digital transformation journey.

The first step in this technical evaluation is to define clear, measurable criteria that align with the company's **short-term operational goals** and **long-term strategic vision**. These criteria should encompass not only the technical capabilities of the alternative solutions but also how these capabilities contribute to achieving broader business objectives. For example, while performance and reliability are important for meeting immediate operational needs, factors such as **scalability**, **security**, **cost-effectiveness**, and **support for emerging technologies** are essential for long-term sustainability. Let's delve into the core elements that should guide this evaluation.

Please refer to Figure 4-1, which explains the 5-step technical evaluation criteria for any potential vendors.

Figure 4-1. *Five-step technical evaluation criteria for vendors*

Performance and scalability: When exploring alternatives to VMware, organizations must focus on performance capabilities – CPU, memory, I/O operations, and the ability to scale as business demands grow. Performance benchmarks should align with the company's immediate workload demands while ensuring enough capacity for future expansion.

For example, Nutanix is a possible technological alternative. Companies must assess its horizontal scalability by adding more nodes to the cluster and vertical scalability by upgrading existing nodes to ensure seamless growth. OpenStack, another contender, offers tailored scalability due to its modular architecture, making it adaptable to a variety of workloads. This allows businesses to dynamically adjust resources like compute, storage, and network as needed.

Evaluating how these platforms handle resource-intensive tasks is critical. For instance, performance metrics such as I/O operations per second (IOPS) or memory management during peak loads can highlight which alternative would best support current and future operations.

Compatibility and interoperability: Compatibility with existing infrastructure is vital when transitioning from VMware. Businesses should ensure that the new solution integrates seamlessly with their hardware, operating systems, and applications. Additionally, interoperability with third-party software from other vendors is essential to avoid operational silos.

For example, Proxmox VE (Virtualization Engine) can manage both virtual machines and containers, offering flexibility, but its integration with storage solutions like Ceph or file systems such as ZFS must be scrutinized. XenServer, on the other hand, needs to be evaluated for its interoperability with existing enterprise solutions like Active Directory and its ability to support various guest operating systems. These considerations help in maintaining a cohesive, multi-vendor IT environment without disrupting operations.

Cost and licensing models: One of the key motivations for reducing VMware dependence is the potential for cost savings. A thorough analysis of alternative licensing models is essential. Organizations must consider initial licensing fees, long-term maintenance costs, and any savings that can be realized through reduced reliance on VMware.

For example, Nutanix's solution may help eliminate VMware licensing costs and can offer substantial savings, but initial hardware investments and ongoing software fees need to be factored into the equation. Similarly, Red Hat Virtualization, with its open-source licensing model, can provide a cost-effective alternative. Companies should compare different licensing options – such as Nutanix's term-based vs. capacity-based licensing – to understand the long-term financial implications and potential ROI.

Support and maintenance: Vendor support is a cornerstone of any transition. It is important to evaluate the quality of support services, including service-level agreements (SLAs), response times, and the availability of patches and updates. For instance, Canonical's support for Ubuntu-based private clouds or OpenStack should be analyzed for SLA performance and its track record in minimizing downtime. A robust support infrastructure can significantly reduce risks during migration and ensure that any post-migration issues are resolved swiftly, minimizing operational disruptions.

Migration complexity: The complexity of migrating workloads from VMware to an alternative solution can be a major hurdle. Companies need to assess the migration tools provided by the vendor, the potential need for reconfiguring applications, and the anticipated downtime during the process.

For instance, Nutanix Move, a migration tool specifically designed for transitioning VMware workloads to Nutanix, offers step-by-step guidance, reducing both complexity and downtime. Similarly, Proxmox VE's built-in migration tools ensure smooth transfers of virtual machines with minimal reconfiguration. Evaluating these tools early in the decision-making process can help prevent costly disruptions during the migration phase.

Reducing VMware dependence is not merely about replacing one technology with another – it's about building a future-ready, scalable, and cost-efficient IT infrastructure. By conducting a meticulous technical evaluation of performance, compatibility, cost, support, and migration complexity, organizations can confidently select an alternative that ensures operational continuity, long-term growth, and reduced vendor lock-in.

4.2 Establishing Clear Migration Paths

Transitioning from VMware to an alternative platform represents a pivotal and strategic choice for any organization. Due to the intricate nature of this process and the significant dependence on VMware within numerous enterprises, it is crucial to establish clear migration pathways to facilitate a seamless, efficient, and successful transition. An effectively designed migration strategy not only reduces operational interruptions but also addresses risks related to downtime, data loss, and potential security threats. Furthermore, a robust migration framework lays the groundwork for future expansion and scalability, enabling the organization to embrace new technologies and evolving business models.

The primary objective of defining clear migration pathways is to create a comprehensive roadmap that directs each phase of the transition, ensuring that the organization's technical and operational needs are fulfilled. An appropriate migration strategy should align with the overarching goals of the business – whether aimed at cost reduction, performance improvement, or strategic innovation – allowing for a transition that is as frictionless as possible. This process entails assessing infrastructure compatibility, prioritizing workloads, and managing resource allocation while also considering timelines and organizational change management to prepare employees effectively.

Organizations typically choose from three primary migration strategies: **Lift and Shift**, **Re-platforming**, and **Refactoring**. Each of these approaches has its own set of advantages, trade-offs, and challenges. Deciding which path to pursue depends largely on the organization's current technical environment, business goals, and long-term IT strategy. Below, we explore these approaches in greater detail, highlighting how they help manage VMware dependence while optimizing for performance, scalability, and agility. Let's delve more into each strategy:

4.2.1 Lift and Shift

The "Lift and Shift" strategy is frequently regarded as the most expedient and least disruptive method for transitioning from VMware to a different platform. This approach entails the direct transfer of applications and workloads to the new environment with minimal alterations, effectively mirroring the existing infrastructure on the new platform. By doing so, it accelerates the migration process and minimizes downtime. Nevertheless, this strategy may not fully enhance performance or leverage the advanced features available in the new platform. Despite this limitation, it remains the preferred choice for many organizations due to the minimal re-engineering required for their workloads. For instance, when migrating from VMware to Nutanix AHV using Nutanix Move, businesses can seamlessly transfer virtual machines (VMs) with only slight modifications. This method preserves the current application architecture, ensuring operational continuity during the transition to Nutanix's hyperconverged infrastructure. Likewise, XenServer (Citrix Hypervisor) provides integrated tools for direct VM migration from VMware, facilitating a smooth transition with minimal adjustments. These direct transfer techniques are particularly attractive to organizations aiming for a swift migration with limited reconfiguration. However, while the Lift and Shift approach is effective for a rapid transition, it often results in migrated workloads that are not fully optimized for the unique advantages of the new environment, potentially overlooking opportunities for performance enhancements and cost savings.

4.2.2 Re-platforming

Re-platforming represents a strategic approach that balances the speed of migration with the need for optimization, necessitating only minor adjustments to applications to align them more effectively with the new platform's capabilities. This method enables organizations to enhance performance, increase scalability, and utilize certain features of the new platform without the need for extensive redevelopment. For instance, when transitioning applications from VMware to OpenStack, re-platforming may require slight modifications to leverage OpenStack's inherent services, such as Heat for orchestration and Neutron for networking. This strategy not only boosts performance but also capitalizes on OpenStack's modular architecture, allowing for adaptation to changing business requirements without a complete redesign of the application framework.

In a similar vein, migrating applications from VMware to Proxmox VE may necessitate optimizing the use of KVM (kernel-based virtual machine) for virtualization and LXC (Linux Containers) for containerization. This ensures that applications are effectively tailored to Proxmox's capabilities, resulting in enhanced efficiency and scalability over time. Re-platforming thus enables organizations to achieve a balance between rapid migration and the positioning of applications to fully leverage the advantages of their new environment.

The re-platforming strategy for transitioning VMware workloads to a public cloud platform, such as Microsoft Azure, often extends beyond the basic Lift and Shift method. It involves making slight modifications to applications or infrastructure to better integrate with Azure's native services. This approach focuses on optimizing workloads for Azure without necessitating a complete overhaul of the underlying architecture, thus enabling organizations to utilize Azure-specific features like Azure SQL Database, Azure App Services, and Azure Kubernetes Service (AKS). In a scenario where VMware workloads are re-platformed to Azure, the workloads are generally migrated to Azure Virtual Machines (VMs) with necessary adjustments to take advantage of Platform as a Service (PaaS) and Infrastructure as a Service (IaaS) offerings. This enhances performance, scalability, and cost-effectiveness. For instance, databases that operate on VMware can be transitioned to Azure SQL, benefiting from managed services that lower operational burdens while improving reliability. This strategy strikes a balance between achieving immediate cloud advantages and pursuing long-term modernization, enabling organizations to leverage cloud-native services while preserving essential application functionality.

4.2.3 Refactoring

For organizations looking to maximize performance and fully exploit the capabilities of the new platform, refactoring offers the most comprehensive solution. Refactoring involves **redesigning** or **re-architecting applications** to take full advantage of the features and scalability of the new platform. While this approach is the most resource-intensive, it also delivers the highest potential gains in performance, agility, and cost efficiency.

Take, for instance, refactoring applications to a microservices architecture on Docker containers offers organizations a scalable and flexible approach to software development. This process involves breaking down monolithic applications into smaller, self-contained services that can be developed, deployed, and maintained independently. By leveraging OpenStack's robust private cloud infrastructure, businesses can easily orchestrate and manage these microservices, ensuring they are resilient and can scale according to demand. This transition not only enhances deployment speed and efficiency but also allows teams to adopt agile practices, fostering a culture of rapid innovation. Additionally, the inherent capabilities of OpenStack, such as its support for containerization and orchestration tools, streamline the integration of continuous delivery and DevOps methodologies, ultimately leading to increased operational efficiency and improved customer satisfaction.

Similarly, organizations refactoring applications for Nutanix AHV (Nutanix Acropolis Hypervisor) can leverage Nutanix's centralized management through Prism and its automation tools to streamline operations. By redesigning applications to fully integrate with AHV's advanced features, businesses can unlock new levels of performance, reduce manual intervention, and ensure their infrastructure is future ready.

The Refactoring approach for migrating VMware workloads to a public cloud platform such as Microsoft Azure involves significantly redesigning and re-architecting applications to fully exploit the capabilities of Azure's native cloud services. This method requires a deeper overhaul of the application's structure, moving away from traditional VMware-based infrastructure to adopt modern cloud-native architectures such as microservices or serverless computing. Refactoring allows organizations to reimagine applications using services like Azure Functions, Azure Logic Apps, or Azure Cosmos DB, thereby enhancing performance, scalability, and resilience. In this scenario, legacy applications may be broken down into smaller, independently deployable components, which can then leverage Kubernetes or containers through Azure Kubernetes Service (AKS). This approach demands more upfront effort and investment but yields substantial long-term benefits, including greater flexibility, reduced technical debt, and improved alignment with agile development methodologies. By embracing refactoring, organizations future-proof their applications for continuous innovation and cloud-native scalability.

4.2.4 Choosing the Right Path and Developing a Roadmap

When evaluating alternatives to VMware and managing dependence on its infrastructure, organizations face a critical decision regarding the approach to migrating their workloads to the cloud or a different on-premises solution. The three primary strategies – Lift & Shift, Re-platforming, and Refactoring – each offer distinct advantages and challenges that can significantly impact an organization's operations and future scalability. Lift & Shift, often the quickest route, enables businesses to transfer applications as-is to an alternative environment, minimizing initial deployment time but potentially sacrificing optimization and destination platform-native benefits. In contrast, Re-platforming allows for some degree of modification to leverage the destination platform-native capabilities, striking a balance between effort and enhancement, yet it may still not fully exploit all the advantages of a destination platform architecture. On the other hand, Refactoring requires a deeper, more comprehensive overhaul of applications to fully align with the destination solution's best practices, leading to improved performance and scalability in the long run, albeit at a greater upfront investment of time and resources. Therefore, choosing the right path necessitates a thorough assessment of the organization's goals, existing workload characteristics, and long-term strategy to ensure an effective transition that minimizes disruptions while maximizing potential gains. Therefore, the choice between Lift and Shift, Re-platforming, and Refactoring depends largely on an organization's current infrastructure, operational goals, and future scalability needs.

Regardless of the migration strategy, establishing a clear roadmap is critical. This roadmap should outline each step of the migration, including technical assessments, infrastructure audits, testing phases, and support requirements. Companies should also allocate time and resources for employee training, ensuring that teams are fully prepared to manage and maintain the new environment post-migration.

By defining clear migration paths, businesses can minimize disruption, reduce the risk of downtime, and position themselves to fully capitalize on their new, more flexible IT infrastructure. Whether the goal is a quick transition, improved performance, or complete re-architecture, each migration path offers a distinct set of benefits and challenges. Through strategic planning, organizations can successfully reduce VMware dependence and pave the way for a more dynamic, scalable, and cost-efficient future.

4.3 Establishing a Task Force and Crafting an Action Plan

Embarking on the journey to reduce VMware dependence and transition to a new platform is a complex, multifaceted undertaking that involves far more than just technical migration. It requires a deep understanding of the organization's strategic goals, resource allocation, and long-term vision. To ensure that this transition is executed seamlessly and with minimal disruption, it is crucial to establish a dedicated task force that will act as the backbone of the entire initiative.

This task force must consist of cross-functional experts from various departments, including IT, operations, finance, risk management, and cybersecurity. Each member of the team will bring a unique perspective and skill set, contributing to a holistic approach that considers all dimensions of the migration. Their role will go beyond managing technical details, as they will also be responsible for addressing operational efficiency, financial implications, governance, compliance, and maintaining high levels of security throughout the process.

A well-structured task force will act as the **organizational glue** that binds together every element of the migration. They will be responsible for conducting a thorough risk assessment, identifying potential roadblocks, and crafting a detailed action plan that includes timelines, resource requirements, and contingencies. This action plan should be a living document, regularly updated as the project progresses, ensuring agility and responsiveness to unforeseen challenges.

Moreover, the task force must ensure that the migration strategy aligns with broader business objectives, such as improving scalability, enhancing performance, and reducing total cost of ownership (TCO). They must also foster collaboration and communication between internal stakeholders, external vendors, and any third-party consultants involved in the migration. Establishing clear roles, responsibilities, and communication channels is vital to prevent misalignment and ensure accountability at every stage.

In crafting an action plan, the task force will need to break down the migration into manageable phases, each with specific milestones and deliverables. These phases might include **initial assessment and planning**, **infrastructure readiness**, **pilot testing**, and the eventual full-scale deployment. By doing so, the team can maintain a focus on both short-term wins and long-term goals, ensuring that the migration progresses smoothly and without compromising business continuity.

Ultimately, this dedicated task force is not just a tactical team but a strategic enabler, ensuring that the migration not only meets technical and operational needs but also drives innovation, agility, and competitive advantage. Through careful planning, rigorous execution, and continuous optimization, the organization will be well-positioned to thrive in a post-VMware environment, leveraging the new platform to its fullest potential.

4.3.1 Forming the Task Force

The formation of a specialized task force is the first critical step in managing the migration. This team should be cross-functional, drawing expertise from a variety of departments to ensure every aspect of the migration is accounted for – from the granular technical details to the broader business impact. The composition of this team is key to creating a comprehensive migration strategy that touches on all necessary elements. Please refer Figure 4-2 consist of Composition, Leadership, and Responsibilities. The Composition section lists IT (Technical Expertise), Operations (Workflow Integration), Security (Address Vulnerabilities), Finance (Budget Management), and Business Units (Ensure Alignment). The Leadership section includes appointing a project manager, experience in large-scale IT projects, coordinating efforts, managing timelines, and maintaining communication. The Responsibilities section highlights defining roles and responsibilities, and scheduling regular meetings and status updates. Each section is represented by an arrow pointing right.

At the heart of the team are stakeholders from IT, operations, security, finance, and key business units. Each of these departments brings a unique perspective to the table. The IT team will naturally provide the technical expertise required for evaluating and implementing alternative platforms, such as Nutanix AHV, OpenStack, or Proxmox VE. They will be responsible for the hands-on work of migration, including configuring systems and managing data transfer. The operations team, on the other hand, ensures that the workflow of the organization remains uninterrupted during the migration, maintaining business continuity throughout the transition.

The security team plays a critical role in identifying potential vulnerabilities that may arise during the migration. They must work proactively to address concerns such as data integrity and compliance, ensuring that sensitive information is safeguarded throughout the process. Meanwhile, the finance department will manage the budget, monitoring expenditures and analyzing cost-saving opportunities that come with reduced reliance on VMware. Finally, representatives from business units ensure that the migration aligns with strategic goals, keeping customer needs and business outcomes at the forefront.They are crucial in the success of the migration as they are able to provide the necessary guardrails for a smooth migration with a minimum of interruption in business processes and applications.

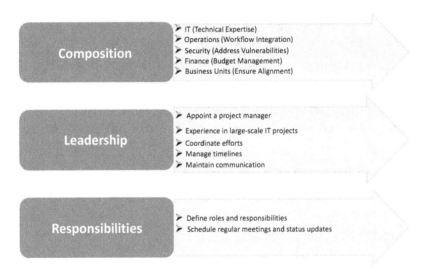

Figure 4-2. *Forming the task force*

This diversity in representation creates a holistic approach to migration planning. Each department brings its expertise to the table, ensuring that no aspect of the migration is overlooked.

4.3.1.1 Leadership and Coordination

No task force can function without strong leadership. Appointing a seasoned project manager to steer the team is crucial to the success of the migration. The project manager, ideally someone with experience leading large-scale IT transitions, such as migrating a financial application from VMware to Nutanix AHV, will be responsible for coordinating efforts across departments. They act as the central hub, managing timelines, ensuring deliverables are met, and keeping clear lines of communication open between team members and executive stakeholders.

Their role is also to keep the project on track, managing any risks or delays that may arise and making swift, informed decisions to maintain momentum. In large migrations, it is easy for tasks to become siloed or for communication breakdowns to cause delays. A strong leader mitigates these risks by fostering collaboration and ensuring that every department is aligned on the project's objectives and timelines.

4.3.1.2 Defining Roles and Responsibilities

Once the task force is formed, clearly defined roles and responsibilities must be established for each member. This clarity is vital for maintaining accountability and ensuring that every aspect of the migration has a designated owner. Each team member should have a specific role that aligns with their expertise.

For example, the IT lead will oversee the technical intricacies of the migration to a platform like Proxmox VE, ensuring compatibility with existing systems, managing the transfer of virtual machines, and troubleshooting any technical challenges that arise. The operations lead will focus on ensuring that day-to-day business activities remain unaffected by the migration, working closely with IT to avoid disruptions. The security lead will be tasked with assessing and mitigating risks, ensuring compliance with industry standards, and implementing security protocols during and after the transition. Finally, the finance lead will track the budget, ensuring cost-effectiveness while managing the financial implications of licensing, hardware investments, and ongoing operational costs.

Regular meetings and status updates are essential for maintaining progress and addressing any issues in real time. By fostering consistent communication, the team can quickly identify and resolve potential bottlenecks, ensuring that the migration remains on schedule.

4.3.1.3 Crafting the Action Plan

With the task force established and roles clearly defined, the next step is to craft a comprehensive action plan. This plan serves as the roadmap for the migration, outlining every phase from initial evaluation to post-migration optimization. It should include key milestones, deadlines, and deliverables for each stage of the project.

The action plan should also be adaptable, allowing for adjustments as the migration unfolds. For example, unexpected technical challenges, such as compatibility issues between VMware and Nutanix AHV, may require the team to revisit certain steps or allocate additional resources. A flexible action plan enables the team to pivot when necessary, without derailing the project.

At its core, the action plan must align with the company's strategic objectives, ensuring that the migration not only reduces VMware dependence but also enhances overall operational efficiency, security, and scalability.

4.3.1.4 Fostering Collaboration and Communication

Successful migrations require seamless collaboration across departments and clear, consistent communication with all stakeholders. The task force must work closely together, leveraging each department's expertise to solve problems and achieve shared goals. Additionally, it is crucial to maintain open lines of communication with the company's leadership team, providing regular updates on progress, risks, and achievements.

The project manager plays a key role in facilitating this communication, ensuring that leadership is kept informed of major milestones and any challenges that arise. This transparency is critical for maintaining support and momentum from executive stakeholders, ensuring that the migration remains a priority within the organization.

4.3.1.5 A Unified Approach to Success

By establishing a dedicated task force and crafting a detailed action plan, organizations can navigate the complexities of migrating away from VMware with confidence. This unified approach ensures that every department is engaged, every potential risk is accounted for, and the migration is executed smoothly and strategically. With strong leadership, clear responsibilities, and a flexible, forward-thinking plan, companies can successfully transition to a more flexible, scalable IT environment, positioning themselves for long-term growth and innovation.

4.3.2 Crafting the Action Plan

Transitioning from VMware to alternative platforms such as Nutanix AHV, Proxmox VE, Red Hat Virtualization, or Microsoft Hyper-V is a strategic move. However, the migration process is complex and requires a well-structured, comprehensive action plan to ensure success. The journey isn't just about switching platforms; it's about creating a future-ready IT environment that is agile, scalable, and cost-effective. Please refer flowchart in Figure 4-3 which represents "Assessment Phase," "Planning Phase," "Execution Phase," "Validation Phase," and "Optimization Phase." The chart visually represents a structured approach to project management or process improvement.

Figure 4-3. Crafting the action plan

4.3.2.1 Assessment Phase: Understanding the Terrain

Like any successful expedition, the migration journey begins with a thorough assessment of the existing VMware environment. This is the "discovery" phase, where the IT team takes a detailed inventory of critical workloads, dependencies, and any potential roadblocks that may hinder the migration. The goal here is to get a clear picture of what is currently running and how it's performing.

For instance, the organization should benchmark key performance metrics – CPU utilization, memory usage, and I/O operations – across critical workloads. These metrics will serve as the baseline to compare alternative platforms such as Red Hat Virtualization or Microsoft Hyper-V. Each platform comes with its unique advantages, but understanding how they stack up against VMware in terms of performance and cost efficiency is essential. Assessing compatibility with existing hardware and infrastructure is also crucial. For example, Nutanix AHV's hyper-converged architecture may offer greater simplicity and cost savings, but only if it aligns with the existing environment.

4.3.2.2 Planning Phase: Charting the Course

Once the lay of the land is clear, the next step is crafting a detailed migration plan. Think of this as drawing the map for the journey. Every detail, from timelines and milestones to resource allocation and potential risks, must be meticulously outlined. Planning should also include mitigation strategies for anticipated challenges – whether it's ensuring application compatibility on XenServer or preparing staff for using Microsoft Hyper-V.

One critical aspect of this phase is training. Migrating from VMware to Microsoft Hyper-V, for instance, requires the IT team to be well-versed in Hyper-V's features, architecture, and management tools. Ensuring your team is up to speed on the new platform minimizes downtime and operational disruptions. Risk mitigation is another key factor here. What if certain applications don't run as smoothly on Nutanix AHV? Identifying these risks early allows the team to devise contingency plans, such as fallback strategies or additional testing phases.

4.3.2.3 Execution Phase: Taking the First Steps

The migration itself should unfold in carefully managed phases, beginning with less critical systems before moving to core workloads. This phased approach allows for testing, adjustment, and optimization of the new platform before fully committing critical workloads. For example, migrating a non-critical web server to Microsoft Hyper-V or Proxmox VE can serve as a testbed for more significant migrations down the line.

Continuous monitoring is the backbone of this phase. As systems are transferred from VMware, it's crucial to keep an eye on performance metrics, ensuring workloads perform optimally in the new environment. A phased execution also provides the opportunity to refine the migration process with each step. For instance, early-stage migration to Nutanix AHV might reveal specific configuration adjustments that improve performance, which can then be applied to later stages of the migration.

4.3.2.4 Validation Phase: Ensuring the Path Is Clear

After the migration of each system, it's time for rigorous validation. This is where the team ensures that everything operates as intended in the new environment. Performance, security, and compatibility with existing systems are scrutinized, and any issues identified are swiftly addressed.

Consider this a hyper-care phase where the dedicated task force works in tandem with the day-to-day operations team. They can identify the applications that ran on VMware function seamlessly on the desired destination solution, for example, Red Hat Virtualization or Microsoft Hyper-V. Do they meet the same security standards? Is the integration with existing storage, networking, and other resources as smooth as expected? Validation should involve not just the IT team but also the stakeholders from business units to ensure the new environment meets their operational needs. During this phase, the organization may uncover optimization opportunities. The idea is to learn as it provides better chances to tweak and fine-tune before the hosting of the next migration wave to the destination platform.

4.3.2.5 Optimization Phase: Reaping the Rewards

With the migration completed and validated, the final phase is optimization. Now, the focus shifts from transitioning to maximizing the new platform's potential. This is where organizations can really leverage the advantages of a destination platform such as Nutanix's centralized management or Proxmox's lightweight infrastructure or Microsoft Hyper-V's deep integration with Windows systems.

A post-migration review is essential to capture lessons learned. What went well? What could have been smoother? This reflection is crucial for future projects and helps ensure continuous improvement in IT processes. The optimization phase is also an opportunity to streamline configurations, fine-tune performance settings, and explore advanced features of the new platform. For example, Nutanix Prism provides powerful analytics and automation tools that can be harnessed to reduce administrative overhead and improve system performance. Additionally, optimizing configurations to suit future needs can include leveraging hybrid cloud capabilities, integrating with public cloud platforms like Microsoft Azure when migrating to Hyper-V technology, or optimizing resource allocation across data centers when using Red Hat Virtualization.

A well-executed migration from VMware to alternative platforms like Nutanix AHV, Microsoft Hyper-V, or Proxmox VE is more than a technical undertaking. It is a strategic transformation that enables companies to break free from vendor lock-in, reduce costs, and future-proof their infrastructure. The key lies in a robust action plan that carefully guides each phase of the migration process – from initial assessment and planning to execution, validation, and optimization. By forming a dedicated task force, aligning cross-functional expertise, and maintaining a steady focus on business objectives, organizations can ensure their migration journey is smooth, strategic, and successful.

4.4 Executing Workload Migration Strategies and Mitigating Risks

Effective workload migration strategies are paramount for minimizing risks and ensuring a seamless transition. Understanding and implementing these strategies can significantly enhance the chances of a successful migration.

One such strategy is **incremental migration**. This approach involves migrating workloads in small, manageable batches, allowing for continuous assessment and adjustments. This method reduces risk by focusing on smaller segments, making it easier to identify and resolve issues as they arise. For instance, an insurance company might start by migrating non-critical customer support applications from VMware to

Nutanix AHV. By beginning with these less critical systems, the IT team can fine-tune the migration process, ensuring that more critical systems transition smoothly later.

Another crucial strategy is **maintaining a hybrid environment** during the transition. This involves running workloads on both the legacy platform and the new platform simultaneously. By doing so, organizations can ensure continuity of operations and minimize disruption. For example, a manufacturing company might operate a hybrid environment, keeping critical production workloads on VMware while transitioning less critical systems, such as HR and finance applications, to OpenStack. This approach ensures minimal disruption to core manufacturing operations while allowing for gradual migration.

Pilot projects are also an effective way to validate migration approaches. By starting with pilot projects, organizations can identify potential issues and refine their strategies before full-scale implementation. For example, a retail company might begin with a pilot project by migrating its inventory management system to Proxmox VE. This non-critical application serves as a test bed for the migration process, allowing the team to resolve any issues and ensure a smoother process for subsequent, more critical migrations.

Effective workload migration strategies are essential for minimizing risks and ensuring a smooth transition. Here is the high-level summary of the strategies we discussed in this section. This is to ensure successful workload migration:

Incremental migration

- **Rationale**: Migrating workloads in small, manageable batches allows for continuous assessment and adjustments, reducing overall risk. This approach is particularly beneficial for complex environments where a full-scale migration might be impractical or too risky.

- **Implementation**:

  - **Plan**: Develop a detailed migration plan that outlines the specific workloads to be migrated in each batch. Prioritize less critical workloads initially to build confidence and refine processes.

  - **Execution**: Begin migrating workloads, for example, from VMware to Nutanix AHV in small batches. This phased approach allows for troubleshooting and optimization at each stage.

  - **Assessment**: After each batch, evaluate performance, identify any issues, and make necessary adjustments before proceeding with the next batch.

- **Example**: An insurance company begins migrating its customer support applications from VMware to Nutanix AHV. By starting with non-critical support tools, the IT team can fine-tune the migration process, ensuring that critical systems are migrated smoothly later on.

Hybrid approach

- **Rationale**: Maintaining a hybrid environment during the transition period ensures continuity of operations and allows for gradual migration. This minimizes disruption and provides a fallback option if issues arise.

- **Implementation**

  - **Assessment**: Determine which workloads are best suited to remain temporarily on the legacy platform (e.g., VMware) and which can be moved to the new platform (e.g., OpenStack) early on.

173

- **Integration**: Set up network and application integrations to ensure seamless communication between workloads running on both platforms.

- **Transition**: Gradually migrate workloads, for example, from VMware to OpenStack, ensuring each transition is smooth and well-documented.

- **Example**: A manufacturing company operates a hybrid environment during its transition to OpenStack. Critical production workloads remain on VMware, while less critical workloads, like HR and finance applications, are moved to OpenStack. This ensures minimal disruption to core manufacturing operations.

Pilot projects

- **Rationale**: Starting with pilot projects allows for validation of the migration approach in a controlled environment. Pilot projects help identify potential issues and enable the refinement of strategies before full-scale implementation, thereby mitigating risks.

- **Implementation**

  - **Selection**: Choose non-critical or less complex applications for the initial pilot migration projects.

  - **Execution**: Migrate these selected applications to a new platform, such as Proxmox VE, closely monitoring the process.

  - **Evaluation**: Assess the success of the pilot migrations, document lessons learned, and refine the migration strategy based on these insights.

- **Example:** A retail company starts with a pilot project by migrating its inventory management system to Proxmox VE. This non-critical application serves as a test bed for the migration process. The team identifies and resolves any issues, ensuring a smoother process for subsequent, more critical migrations.

In addition to these strategies, mitigating migration risks is crucial to ensure data integrity and minimize operational disruptions. Protecting data integrity is paramount during migration. Implementing robust data backup and recovery plans can safeguard against data loss. For example, a technology firm migrating to Red Hat Virtualization might develop comprehensive data backup and recovery plans, using tools like rsync with checksum verification to ensure data integrity. Employing data validation tools further verifies the accuracy and completeness of the data post-migration.

Minimizing downtime is another critical aspect. Scheduling migrations during off-peak hours can significantly reduce the impact on operations. For instance, a healthcare provider might schedule its migration to XenServer during weekends and late-night hours when patient activity is at its lowest. Having a rapid rollback plan in place allows for a quick reversion to the original environment if any issues arise, ensuring minimal operational disruption.

Continuous performance monitoring throughout the migration process is essential to identify and address any performance issues promptly. Establishing performance baselines before migration provides a reference point for post-migration performance. An e-commerce company migrating to Nutanix AHV might use monitoring tools like Nagios and Prometheus to track performance metrics continuously. By comparing these metrics with pre-migration baselines, they can quickly identify and resolve any performance issues, ensuring a smooth transition and maintaining high operational standards.

Thorough testing in a staging environment that closely mirrors the production environment is crucial. This includes functional testing, load testing, and user acceptance testing to identify and resolve potential issues before the actual migration. Clear and open communication with all stakeholders throughout the migration process helps manage expectations and ensures everyone is prepared for potential disruptions. Ensuring the migration team is skilled and experienced with the platforms involved, along with keeping detailed documentation of the migration process, can significantly reduce the risk of errors and aid in troubleshooting and future migrations.

4.5 Overcoming Integration Challenges and Ensuring Compatibility

4.5.1 Strategies for Overcoming Integration Challenges

Integration challenges are common when incorporating a new platform into an existing IT ecosystem. Overcoming these challenges is crucial for a successful transition.

> **Standardization:** Adopt industry standards and protocols to facilitate interoperability between systems. Standard APIs and data formats can simplify integration and ensure consistency. Adopt industry standards and protocols to facilitate interoperability between Microsoft Hyper-V, OpenStack, and existing systems. Standard APIs and data formats can simplify integration and ensure consistency.

Middleware solutions: Utilize middleware to bridge gaps between incompatible systems. Middleware solutions can enable seamless communication and data exchange, ensuring operational continuity. Utilize middleware to bridge gaps between incompatible systems, such as connecting legacy applications with Proxmox VE or Microsoft Hyper-V. Middleware solutions enable seamless communication and data exchange, ensuring operational continuity.

Custom integration: Develop custom integration solutions when necessary. Leverage APIs and other tools to connect disparate systems, ensuring they work together effectively. Custom integrations should be well-documented and tested thoroughly. Develop custom integration solutions when necessary. Leverage APIs to connect Nutanix AHV or Microsoft Hyper-V with other systems, ensuring they work together effectively. Custom integrations should be well-documented and thoroughly tested.

4.5.2 Ensuring Compatibility

Ensuring compatibility between new and existing systems is vital to avoid operational disruptions.

Thorough testing: Conduct comprehensive testing of all integrated systems. Testing environments should mirror the production setup to accurately identify potential issues. Regular testing cycles ensure ongoing compatibility. Conduct

177

comprehensive testing of all integrated systems. For instance, create a test environment that mirrors the production setup for Red Hat Virtualization or Microsoft Hyper-V to accurately identify potential issues. Regular testing cycles ensure ongoing compatibility.

Vendor collaboration: Work closely with vendors to ensure their solutions are compatible with your existing infrastructure. Utilize vendor resources and expertise to address any compatibility issues that arise during the migration process. Work closely with vendors like Canonical for OpenStack or Microsoft to ensure their solutions are compatible with your existing infrastructure. Utilize vendor resources and expertise to address any compatibility issues that arise during the migration process.

Ongoing support: Establish ongoing support agreements with vendors to address any post-migration issues. Continuous support ensures that compatibility issues are promptly identified and resolved. Establish ongoing support agreements with vendors like Mirantis for OpenStack, Microsoft for Hyper-V, or Azure to address any post-migration issues. Continuous support ensures that compatibility issues are promptly identified and resolved.

4.6 Mitigating Vendor Lock-In Risks Through Interoperability

Vendor lock-in can significantly limit an organization's flexibility and increase long-term risks. To mitigate these risks, adopting robust interoperability strategies is essential. Here are several effective approaches:

Open standards: Utilizing open standards helps ensure compatibility and flexibility, reducing dependency on any single vendor. Open standards are publicly available and provide a common framework that different vendors can adopt, ensuring that systems can work together seamlessly. Opt for vendors that support open standards like OpenStack, Proxmox VE, or Microsoft Hyper-V. These standards facilitate interoperability and future-proof your IT infrastructure. Avoid proprietary technologies that limit integration with other systems.

Multi-vendor environments: A multi-vendor strategy can foster competition, increase bargaining power, and prevent over-reliance on a single provider. By diversifying vendors, organizations can compare offerings and leverage better deals. Integrate different solutions such as Nutanix AHV alongside existing systems. This approach ensures a more resilient and flexible IT environment. It also allows for cross-training IT staff on multiple platforms, reducing the risk associated with staff turnover.

179

For example, a healthcare organization deploys Nutanix AHV for its virtualized environment while maintaining some legacy systems on VMware. This strategy allows the organization to avoid being locked into a single vendor's ecosystem and gives them the flexibility to choose the best technology for different workloads. When negotiating contracts, they can leverage the presence of multiple vendors to secure better terms.

Containerization: Containerization decouples applications from the underlying infrastructure and the operating system, enhancing portability and easing the transition between different environments. Containers provide a consistent runtime environment, which makes it easier to move applications across different cloud providers or on-premises infrastructure. Employ containerization technologies like Docker and Kubernetes within platforms such as Red Hat Virtualization and Microsoft Hyper-V. This method ensures that applications remain portable and flexible, regardless of the underlying infrastructure.

For example, an e-commerce company uses Docker to containerize its microservices architecture. By orchestrating these containers with Kubernetes, they can easily deploy applications across multiple cloud providers, such as AWS, Google Cloud, and Azure. This setup not only enhances their flexibility and scalability but also provides a robust disaster recovery solution, as they can quickly shift workloads from one cloud provider to another if necessary.

4.7 Multi-vendor Virtualization Environments and Prioritizing Security

Organizations these days are increasingly opting for multi-vendor virtualization environments rather than relying on a single vendor. This approach offers several compelling benefits, including enhanced flexibility, improved resilience, and cost optimization. By leveraging multiple platforms like Microsoft's Hyper-V, Proxmox, and Nutanix, businesses can tailor their virtualization strategy to meet diverse workload requirements while avoiding vendor lock-in. For instance, Hyper-V provides robust integration with Windows environments and enterprise-level features, while Proxmox offers a versatile open-source solution that combines KVM-based virtualization with container support through LXC. Nutanix, on the other hand, excels in hyper-converged infrastructure, seamlessly integrating storage and compute resources, which facilitates rapid scaling and efficient management. By harnessing the strengths of these varied solutions, organizations can optimize resource utilization, streamline operations, and maintain high availability, all while ensuring they are not tied to a single supplier's ecosystem. This multi-vendor strategy not only enhances innovation by allowing companies to take advantage of specialized tools but also fortifies their infrastructure against potential vendor disruptions, making it a modern best practice in IT management.

However, this multifaceted setup can introduce significant security challenges, making it imperative for organizations to prioritize the safeguarding of their virtual environments. With the integration of multiple vendors, potential vulnerabilities can arise from disparate security protocols and management practices, leading to increased risks of data breaches and cyberattacks. Ensuring robust security in such environments requires a comprehensive strategy that encompasses uniform security policies, real-time monitoring, and regular vulnerability assessments across all platforms. By adopting a proactive approach to security, organizations can better protect their assets, maintain compliance with regulatory standards, and foster a resilient infrastructure that adapts to the evolving threat landscape. Prioritizing security within multi-vendor virtualization not only mitigates risks but also enhances overall organizational trust and operational continuity. Let us discuss some of the methodologies to prioritize security in multivendor environments.

Security strategies: Security is paramount in multi-vendor environments, where increased complexity can introduce vulnerabilities. Implementing a cohesive and thorough security strategy ensures that all components, regardless of vendor, are protected and compliant.

Unified security policies: Implementing unified security policies across all platforms and vendors is crucial. Consistent policies ensure comprehensive protection and compliance with regulatory requirements. A fragmented approach can leave gaps in security, leading to vulnerabilities that attackers might exploit. To achieve this, organizations should develop a comprehensive security policy framework that can be applied universally across all platforms, including Nutanix AHV, OpenStack, or Microsoft Hyper-V. This framework should cover all aspects of security, including access controls, data protection, and incident response procedures. Using centralized management tools can help enforce these policies uniformly, reducing the risk of human error and ensuring that all systems adhere to the same security standards.

Zero Trust model: Adopting a Zero Trust security model is essential in today's threat landscape. This model operates on the principle that no entity, whether inside or outside the network, should be trusted by default. Instead, every access request must be verified and authenticated, ensuring that only legitimate users and devices gain access to the network and its resources. In environments using Proxmox VE and Microsoft Hyper-V, this means implementing strict identity verification processes, such as multi-factor authentication (MFA) and continuous monitoring of user activities. By assuming that every entity is potentially hostile, organizations can significantly reduce the risk of breaches caused by compromised credentials or insider threats. Regularly updating and patching systems, as well as employing advanced threat detection and response tools, further enhances security in a Zero Trust model.

Regular audits: Conducting regular security audits and vulnerability assessments is vital for maintaining a secure environment. These audits help identify and address potential weaknesses before they can be exploited by attackers. Regular assessments ensure that security measures remain effective and up to date with evolving threats. This involves scheduling periodic reviews of system configurations, access logs, and security controls. Automated tools can be employed to scan for vulnerabilities and misconfigurations, providing detailed reports that guide remediation efforts. In addition to internal audits, engaging third-party security experts to perform independent assessments can provide valuable insights and help validate the effectiveness of the organization's security posture.

To conclude, organizations must adopt a multi-vendor model to protect themselves from the changing dynamics of the industry and to remain relevant amidst continuously evolving trends. Additionally, it is crucial to emphasize security within a multi-vendor platform, as this approach aids in risk mitigation and bolsters overall organizational trust and operational continuity.

4.7.1 Best Practices for Security in Multi-vendor Environments

Adopting best practices for security is essential to ensuring a robust and resilient IT infrastructure in a multi-vendor setup, especially when integrating platforms like Microsoft Hyper-V, OpenStack, and Nutanix AHV. By following a comprehensive security strategy, organizations can safeguard their systems and data while maintaining flexibility across different virtualization technologies.

> **Encryption:** Strong encryption is the foundation of a secure multi-vendor environment. IT decision-makers must prioritize the encryption of data at rest and in transit to protect sensitive information from unauthorized access. Within environments using Microsoft Hyper-V, OpenStack, and Proxmox VE, organizations should implement TLS (transport layer security) for data in transit and AES (advanced encryption standard) for data at rest. Encryption keys should be managed securely, with regular rotation and stringent access controls. For example, Microsoft Hyper-V supports BitLocker encryption for virtual hard disks, providing an additional layer of security in hybrid environments. Implementing encryption at both the infrastructure

and application levels ensures that data remains protected throughout its entire lifecycle, even across diverse platforms.

Access controls: Strict access control mechanisms are essential to securing multi-vendor environments. Role-based access controls (RBAC) should be implemented to ensure that only authorized personnel can access critical systems. In Microsoft Hyper-V and Nutanix AHV environments, IT teams should create user groups and assign permissions based on job functions. This reduces the risk of unauthorized access to privileged accounts. Automated tools for monitoring and logging access activities can be crucial in detecting anomalies or unauthorized attempts. Regular reviews of access permissions will ensure that as organizational roles evolve, so do access rights, keeping the infrastructure secure without impeding business functions.

Incident response: A comprehensive incident response plan is crucial for managing potential security breaches and ensuring operational resilience. In a multi-vendor ecosystem, such as one incorporating Microsoft Hyper-V and Red Hat Virtualization, organizations need a well-defined plan that outlines roles, responsibilities, and communication protocols. This plan should also include specific procedures for containment, eradication, and recovery. For instance, regular security drills and simulations involving all relevant teams will prepare them for real-world scenarios,

ensuring swift and coordinated responses. After
an incident, post-incident reviews are essential for
identifying gaps in the process and improving future
response strategies.

By implementing these best practices – encryption, access controls,
and incident response – organizations can ensure their multi-vendor
virtualization environments remain secure, scalable, and adaptable. This
approach not only mitigates the risks of vendor lock-in but also enhances
security across diverse platforms, enabling businesses to leverage the
strengths of technologies like Microsoft Hyper-V, Nutanix, and OpenStack
while maintaining control over their IT infrastructure.

4.7.1.1 An Example of Strengthening Unified Security Policies with Microsoft Hyper-V

In light of the aforementioned security strategies, it is essential to
thoroughly examine each vendor and their respective services. Let us
discuss Microsoft Hyper-V within a multi-vendor security framework
and understand how we can strengthen security with this provider.
Integrating Microsoft Hyper-V into a multi-vendor security approach not
only bolsters the overall resilience of the IT infrastructure but also works
synergistically with other platforms such as Nutanix AHV and OpenStack.
By implementing best practices – including unified policies, Zero Trust
frameworks, encryption, and effective incident response – organizations
can establish a secure and robust environment. Hyper-V, with its seamless
integration into Microsoft's extensive security ecosystem, provides distinct
advantages that enhance security not only for Hyper-V workloads but
also across the entire virtualization spectrum. Therefore, examining
the Microsoft Hyper-V platform serves as an illustrative example for
understanding these best practices more effectively.

Integrating Microsoft Hyper-V into a unified security framework calls
for a seamless alignment of policies across all platforms. Like Nutanix

AHV and OpenStack, Hyper-V benefits from comprehensive policy management that covers access control, data protection, and incident response procedures. However, what sets Hyper-V apart is its tight integration with **Microsoft Active Directory (AD)**, a feature that allows organizations to extend existing identity and access management (IAM) strategies. By using AD, organizations can enforce RBAC (role-based access control) with granular permissions, ensuring that administrators and users have precisely the rights they need, without the risk of over-provisioning.

This approach harmonizes well in multi-vendor environments where disparate systems need to operate securely as a cohesive whole. For instance, using AD's centralized identity management across Hyper-V, Nutanix AHV, and OpenStack ensures that organizations maintain consistent identity verification and access rules across platforms.

4.7.1.2 Adapting the Zero Trust Model with Microsoft Hyper-V

When adopting a Zero Trust Model, Hyper-V's compatibility with multi-factor authentication (MFA) and conditional access adds another layer of protection. Microsoft Hyper-V's native integration with Azure Active Directory (Azure AD) enables seamless MFA enforcement across environments. In multi-vendor setups, this can help ensure that access is strictly verified, regardless of whether users are operating in a Hyper-V, Nutanix, or OpenStack instance.

By leveraging Windows Defender Credential Guard, Hyper-V fortifies the environment against credential-based attacks, providing additional security for administrative credentials, often a vulnerable point in virtualization environments. This model can be scaled across a multi-vendor ecosystem, ensuring that user identities, endpoints, and application access are thoroughly validated at every checkpoint, further reducing risks of unauthorized access in environments like OpenStack and Proxmox VE.

187

4.7.1.3 Regular Audits: Microsoft Hyper-V's Compliance Enhancements

In a multi-vendor landscape, regular audits and vulnerability assessments become even more critical, and Hyper-V provides specialized tools that enhance this process. With System Center Virtual Machine Manager (SCVMM), organizations can automate and schedule security assessments, including the review of configurations and adherence to compliance standards. This tool offers deep insights into virtualized workloads, allowing IT teams to cross-check configurations not only for Hyper-V but for all other integrated platforms, ensuring a holistic view of the entire environment.

The auditing process in Hyper-V can also take advantage of Windows Event Logs and Azure Security Center, which provide advanced monitoring capabilities. These tools can be integrated with broader monitoring frameworks used for Nutanix AHV and OpenStack, giving organizations a unified dashboard for tracking potential vulnerabilities and incidents.

4.7.1.4 Encryption: Elevating Data Security in Microsoft Hyper-V

Data encryption is a cornerstone of security across all virtualization platforms, but Microsoft Hyper-V takes it a step further by offering BitLocker encryption for virtual machine (VM) disks. This feature ensures that data, even when at rest, is encrypted and safeguarded against unauthorized access. The integration with Shielded VMs provides additional protection by ensuring that sensitive workloads are encrypted and only run on trusted hosts.

In comparison to other platforms, like Nutanix or OpenStack, Hyper-V's native encryption capabilities simplify the process for organizations using a Windows-based ecosystem. TLS (transport layer security) is already widely used across all platforms for data in transit, but Hyper-V's unique encryption framework ensures that sensitive business data remains protected at multiple levels – whether on-premises or in the cloud.

4.7.1.5 Incident Response: Microsoft Hyper-V's Edge in Swift Recovery

When incidents occur in a multi-vendor environment, an effective incident response plan is essential. Microsoft Hyper-V's deep integration with Azure Backup and Site Recovery ensures swift containment and recovery from security incidents. In case of a breach, Azure Backup provides automated restoration capabilities, while Azure Site Recovery can orchestrate failovers and disaster recovery across hybrid and multi-cloud environments. These capabilities are particularly beneficial when working with other platforms, like Nutanix AHV or Red Hat Virtualization, allowing organizations to minimize downtime and protect critical applications. Incorporating Microsoft Security Center into the incident response workflow brings advanced threat detection and mitigation tools into play. This not only helps with the immediate containment of threats but also provides detailed analytics and post-incident reviews, helping identify the root cause and improving future defense strategies.

4.7.1.6 Cross-Platform Security Orchestration and Monitoring

Employ unified security monitoring tools such as Azure Security Center and Windows Event Logs to orchestrate security across multiple platforms, providing real-time insights into potential threats and vulnerabilities.

Please refer to the below diagram to see a high-level overview of best practices for security in multi-vendor environments.

4.8 Summary

As the digital landscape continues to evolve, businesses must remain agile and innovative to keep up with changing market demands. However, many organizations find themselves deeply entrenched in reliance on a single technology provider, such as VMware, without fully recognizing the associated risks. VMware's long-standing dominance in virtualization has made it a cornerstone of IT infrastructure for data centers and enterprise workloads, but this dependency can present significant challenges to future growth and flexibility.

The over-reliance on VMware exposes organizations to several risks, including rising operational costs, limited innovation, and potential difficulties in adapting to new technologies like multi-cloud and hybrid environments. Vendor lock-in can stifle a company's ability to pivot quickly in response to market changes, creating a bottleneck for scaling and modernizing IT strategies. This chapter encourages decision-makers to critically examine their organization's dependency on VMware, questioning whether this dependence could undermine long-term goals and operational agility.

By evaluating alternative solutions, this chapter offers actionable strategies for minimizing VMware dependence. It stresses the importance of balancing innovation with operational efficiency while maintaining security and flexibility. Key steps include assessing alternative platforms, creating migration roadmaps, and building resilient, multi-vendor ecosystems that reduce reliance on a single provider. These insights are essential for organizations aiming to unlock cost savings, enhance scalability, and future-proof their IT environments against a rapidly shifting technological landscape.

In conclusion, reducing VMware dependence requires careful planning and strategic execution, but the rewards are clear: greater flexibility, increased innovation, and a secure, adaptable IT architecture that can grow alongside the business. This chapter provides the roadmap to navigating these challenges, helping organizations to transform their infrastructures for the future.

CHAPTER 5

Microsoft As an Alternative

One significant aspect we want to cover in this book is to inform readers about the various options available when seeking alternatives to VMware. Given that Microsoft is a prominent leader in the industry, it is essential to include a discussion of Microsoft as a viable alternative. In this chapter, we will begin by analyzing Microsoft's primary product for server virtualization, known as Hyper-V, and explore its historical development over time. Additionally, we will emphasize the critical components of Hyper-V's backend architecture. The latter part of this chapter will focus on the increasingly popular hybrid cloud strategy within the industry. We will introduce Azure Stack HCI (or Azure Local) as a distinctive product that integrates the Microsoft Azure portal and its associated features, serving as a unified management platform for infrastructure resources deployed within your own datacenter. Finally, we will present a comparison between VMware's offering in this area, known as VMware Cloud Foundation (VCF), and Microsoft Azure Stack HCI.

5.1 Deconstructing Microsoft Hyper-V

As organizations progressively adopt virtualization to enhance their IT infrastructures, Microsoft's Hyper-V has become a powerful and dependable option. This platform is specifically engineered to support the

© Sumit Bhatia and Chetan Gabhane 2025
S. Bhatia and C. Gabhane, *Navigating VMware Turmoil in the Broadcom Era*,
https://doi.org/10.1007/979-8-8688-1264-4_5

development and administration of virtual machines (VMs), providing
scalability, efficiency, and a comprehensive set of features that render
it an attractive option for businesses of varying sizes. In this section, we
will analyze Hyper-V, examining its architecture, features, applications,
advantages, and recommended practices for effective implementation.

Hyper-V is a native hypervisor created by Microsoft, designed to
facilitate the creation and management of virtualized environments
on Windows Server as well as on Windows Desktop Pro and Enterprise
editions. This technology enables the abstraction of hardware resources,
permitting the simultaneous operation of multiple operating systems on a
single physical machine, which optimizes resource utilization and reduces
hardware expenses.

5.1.1 The History of Microsoft Hyper-V: A Journey Through Virtualization

In the realm of IT infrastructure, virtualization has fundamentally changed
the manner in which organizations implement and oversee their server
environments. Microsoft Hyper-V, a powerful virtualization platform, has
been instrumental in this evolution. A comprehension of its historical
development provides insight into the larger story of technological
advancement and the transition to cloud computing. This section will
examine the captivating history of Microsoft Hyper-V, highlighting its
beginnings, key milestones, and its emergence as a premier virtualization
solution.

5.1.2 Early Beginnings: The Rise of Virtualization

The idea of virtualization originated in the 1960s with IBM's introduction
of time-sharing systems, which enabled multiple users to concurrently
utilize a single mainframe computer. Despite this early development,

virtualization remained predominantly a specialized and intricate technology, mainly adopted by large corporations and governmental organizations. It was not until the early 2000s that virtualization started to achieve widespread acceptance in the general market.

As businesses started to recognize the value of server consolidation and resource optimization, the need for a comprehensive virtualization solution became apparent. VMware was at the forefront of this movement, offering a robust virtualization platform that captured the attention of IT professionals and businesses. Microsoft, recognizing both the opportunity and the necessity, sought to enter this competitive space and offered Hyper-V first with its Windows 2008 offering. Let us understand more about the journey of Hyper-V from Windows 2008 to the present release of the Windows 2022 platform.

5.1.3 The Birth of Hyper-V: 2008

Microsoft's formal entry into the virtualization market came with the release of Windows Server 2008 on June 26, 2008. This landmark operating system included the first version of Hyper-V, also known as "Viridian" during development. Hyper-V offered a native hypervisor to run virtual machines on x86-64 architecture, allowing businesses to consolidate their physical hardware into virtual environments.

The initial reception of Hyper-V was overwhelmingly favorable. IT administrators valued its smooth integration with Windows Server and Active Directory, as well as its cost-effectiveness and user-friendliness. The launch of Hyper-V alongside Windows 2008 allowed it to establish a distinct position among organizations seeking to enhance their IT infrastructure while avoiding the expenses linked to VMware's products.

The release of Windows Server 2008 R2, along with Hyper-V, marks a pivotal development in virtualization technology, offering organizations a powerful platform for the creation and management of virtualized computing environments. Introduced as part of Microsoft's Windows Server 2008 R2, Hyper-V integrates effectively with the Windows operating system and presents enhanced features that cater to both small and large enterprises. A notable feature of Hyper-V is its capability to support virtual machines (VMs) operating on various operating systems, enabling users to optimize resource utilization by consolidating multiple workloads onto a single physical server. The introduction of functionalities such as Live Migration allows administrators to transfer running VMs between physical hosts without experiencing downtime, thereby significantly enhancing operational flexibility and availability. Additionally, the dynamic memory feature enables Hyper-V to adjust memory allocation to VMs in real-time based on their workload requirements, thereby optimizing performance and efficiency. The management tools provided by Hyper-V are designed to be user-friendly, allowing IT professionals to create, configure, and monitor their virtual environments through a centralized interface. This efficient management is further supported by strong security features, including virtual network isolation and integration with Active Directory, which ensure the protection of sensitive data in a multi-tenant environment. Moreover, Windows Server 2008 R2 Hyper-V accommodates various storage options, such as direct access to iSCSI and Fiber Channel storage, thereby enhancing the scalability of virtualization solutions and allowing businesses to expand their virtual infrastructure as necessary. In summary, Windows Server 2008 R2 Hyper-V not only facilitates the deployment of virtual servers but also offers a scalable and secure foundation that promotes business continuity and meets the increasing demands of contemporary IT environments.

5.1.4 Evolution and Features: 2012 and 2012 R2

The launch of Windows Server 2012 and 2012 R2 brought substantial improvements to Hyper-V. This iteration introduced a variety of advanced functionalities, such as Hyper-V Replica for disaster recovery, a new hard disk format, and network virtualization. Additionally, scalability saw considerable enhancement, allowing for the support of significantly larger virtual machines and physical hosts compared to previous versions.

Windows Server 2012 offers a notable feature in its extensive support for various guest operating systems, facilitating the smooth integration and management of different workloads. The hypervisor-based architecture of Hyper-V guarantees that virtual machines function with optimal performance and security, utilizing advanced capabilities such as dynamic memory and network virtualization to effectively distribute resources according to real-time requirements:

- **Dynamic memory**: This feature allows Hyper-V to allocate memory dynamically to virtual machines based on demand, improving resource management and efficiency.

- **Storage spaces**: This capability allowed for enhanced storage management, enabling users to pool storage resources and create highly available storage solutions.

- **Virtual networking**: Microsoft introduced virtual switches that improved network performance and security for virtual machines.

These enhancements marked an important doubling down on Hyper-V's competitive edge against VMware. By focusing on increased efficiency and cost-effectiveness, Microsoft positioned Hyper-V not just as a strong alternative but as a powerful, enterprise-ready solution.

With the introduction of Windows Server 2012 R2 Hyper-V, Microsoft has significantly enhanced the features available compared to its predecessor. This version brings notable advancements in scalability, performance, and management capabilities. It accommodates an impressive capacity of virtual machines – up to 64 per host and 1 TB of RAM allocated to each virtual machine – making it ideal for organizations of all sizes. A key highlight is the Virtual Machine Generation 2, which utilizes UEFI firmware, facilitating a more secure boot process and quicker boot times, along with the capability to use virtual SCSI adapters, allowing for the addition of virtual hard disks while the virtual machine is operational.

The platform also improves live migration functionalities, enabling the transfer of workloads between hosts without any downtime, which is essential for maintaining business continuity during maintenance or hardware upgrades. Additionally, the introduction of the Hyper-V Replica feature allows organizations to implement disaster recovery solutions with ease, enabling the replication of virtual machines to a secondary site for failover purposes. Enhanced networking support is also included, which improves bandwidth management and increases throughput while offering features such as quality of service (QoS) and VM switch extensions. Moreover, integration with Windows PowerShell allows administrators to automate management tasks, enhancing efficiency and minimizing the risk of errors during repetitive operations. The System Center suite can also be utilized in conjunction with Hyper-V for comprehensive datacenter management, providing IT teams with greater control and visibility over their virtual environments. In conclusion, Windows Server 2012 R2 Hyper-V marks a significant leap forward in virtualization technology, equipping organizations with the necessary tools to innovate and expand their IT infrastructure while ensuring resilience, scalability, and streamlined management.

5.1.5 Expanding Capabilities: Windows Server 2016 and Azure Integration

Windows Server 2016 brought significant advancements to Hyper-V, like support for Docker and containers, nested virtualization, and shielded VMs for improved security. It also received improvements to software-defined networking and storage management capabilities. The add-on, particularly in integration with Microsoft's cloud platform, Azure. Features like

- **Nested virtualization**: This feature allowed users to run Hyper-V inside a virtual machine, paving the way for more complex development and testing environments.

- **Shielded VMs**: This provided enhanced security for virtual machines, protecting them from unauthorized access and improving compliance capabilities. Generation 2 VMs that have a virtual trusted platform module (vTPM), are encrypted with BitLocker, and can only be run on hosts attested and approved by Host Guardian Service (HGS).

- **Storage replica**: This capability allowed for synchronous and asynchronous replication of virtual machines, enhancing disaster recovery options.

The collaboration with Azure represented a pivotal strategic advancement for Microsoft, enabling hybrid cloud solutions that integrated on-premises resources with cloud functionalities. This transition addressed the growing need for scalable and adaptable infrastructure among organizations of varying sizes. The Azure Site Recovery feature, which is built-in, facilitates real-time replication of virtual machines from on-premises Hyper-V hosts to Azure, thereby ensuring business continuity and minimizing downtime during unforeseen failures or disasters. Additionally, the Azure Backup service offers a dependable, cloud-based method for safeguarding essential data, while the Azure Shared Files feature promotes effortless file sharing between on-premises and cloud resources. Hyper-V's capability to leverage Azure Resource Manager templates allows for uniform deployment of virtual machines, and its integration with Azure Active Directory guarantees secure access and identity management across both environments. Moreover, new tools like Azure Migrate aid in evaluating, planning, and executing migrations to Azure, thus streamlining the transition to the cloud. Collectively, these features render Hyper-V on Windows Server 2016 an essential asset for organizations aiming to modernize their infrastructure while leveraging cloud capabilities, ultimately fostering innovation and enhancing operational efficiency. This integration not only boosts IT productivity but also enables businesses to respond more adeptly to market demands by facilitating the swift deployment of applications and services.

5.1.6 Windows 2019 and the Cloud Era

As organizations increasingly transition to hybrid cloud architectures, Hyper-V provides a powerful platform for operating both on-premises and cloud-based virtual machines, facilitating seamless integration and management across varied infrastructures. A notable advantage of Windows Server 2019 Hyper-V is its improved scalability and performance, enabling businesses to effectively manage substantial workloads while optimizing resource utilization, thereby enhancing overall operational efficiency. Additionally, features such as Shielded VMs and the Windows Admin Center contribute to heightened security and streamlined management, addressing significant challenges in the current threat landscape. The orchestration capabilities of Hyper-V align effectively with Microsoft Azure, promoting a unified cloud strategy that allows organizations to capitalize on the advantages of cloud resources, including flexibility, elasticity, and cost efficiency, all while maintaining control over their data. This hybrid model empowers enterprises to swiftly adapt to evolving market conditions, dynamically scale their IT resources, and foster innovation without being hindered by conventional infrastructure constraints. Moreover, Hyper-V's compatibility with a range of operating systems and applications enhances its attractiveness, making it an essential tool for organizations embracing the cloud era. Ultimately, Windows Server 2019 Hyper-V signifies a strategic investment for businesses seeking to fully leverage the benefits of virtualization and cloud computing, enabling them to optimize their current datacenters and prepare for a future where agility and resilience are critical. Windows Server 2019 introduced features tailored for cloud and hybrid cloud environments, with Hyper-V integrating with Azure services such as Azure Backup, Azure File Sync, and the Storage Migration service. It also launched the Windows Admin Center to simplify the management of servers, clusters, and hyper-converged infrastructure.

Microsoft officially announced the discontinuation of the free Hyper-V Server with the release of Windows 2019, marking it as the final edition of this offering. However, support for Hyper-V Server 2019 is scheduled to conclude in 2029. It is important to clarify that this does not signify the termination of the Hyper-V Server itself; rather, it pertains solely to the cessation of the "Free" version of this Microsoft variant. Hyper-V will remain accessible as a feature that can be activated in Windows 2022 and subsequent versions.

5.1.7 The Present and Future: Hyper-V in a Cloud-First World

As we are in 2025, Microsoft Hyper-V continues to advance significantly. It has evolved beyond a mere hypervisor to become a crucial component of Microsoft's overarching cloud strategy. Organizations are increasingly leveraging Hyper-V within Windows Server in conjunction with Azure, which facilitates smooth transitions to cloud environments and hybrid setups.

With ongoing innovations in Hyper-V, Microsoft is dedicated to tackling the security, scalability, and efficiency challenges that organizations encounter in a complex IT landscape. As more organizations embrace cloud-first strategies, current Windows Server 2022 plays a vital role in transforming virtualization, particularly through its integrated

Hyper-V features. This latest version of Windows Server enhances the capabilities of Hyper-V, enabling businesses to establish more resilient, efficient, and scalable virtual environments designed for hybrid cloud implementations.

Key advancements, including secured-core servers, improved Windows Containers, and enhanced management functionalities via Windows Admin Center, empower Hyper-V to provide businesses with the flexibility to operate seamlessly across both on-premises infrastructures and public cloud resources. Organizations can utilize these improvements to dynamically adjust workloads, optimize resource distribution, and implement robust security protocols that safeguard sensitive information in an era marked by increasingly sophisticated cyber threats.

As businesses shift toward hybrid and multi-cloud architectures, the significance of Hyper-V intensifies, serving as a vital link that integrates existing on-premises resources with diverse cloud platforms to foster a fluid and agile IT environment. Additionally, the rising trend of edge computing demands a more distributed virtualization approach, and Windows Server 2022 with Hyper-V is ideally equipped to meet these emerging needs by enabling edge deployments that ensure low latency while maintaining centralized management and orchestration.

As we look to the future, the evolution of cloud services suggests that Hyper-V will play a pivotal role in a cloud-first environment, focusing on intelligent workloads, automation, and AI-driven management solutions. These advancements are expected to not only enhance performance but also facilitate improved decision-making processes. In this context, Microsoft is dedicated to incorporating innovations into Hyper-V that are compatible with cloud-native architectures, thereby solidifying Hyper-V's significance and establishing it as a fundamental component for organizations aiming to fully utilize their IT infrastructure within a hybrid cloud framework. Ultimately, Windows Server 2022 and its Hyper-V functionalities exemplify Microsoft's commitment to empowering enterprises to effectively navigate the challenges of contemporary computing landscapes while optimizing operational efficiency and seizing new opportunities in a digital-first economy.

5.1.8 Architecture of Hyper-V

In the dynamic realm of virtualization technologies, Microsoft Hyper-V has established itself as a powerful solution for enterprise-level virtualization. It is engineered to offer a flexible platform for executing multiple operating systems as virtual machines (VMs) on Windows Server. Hyper-V not only maximizes hardware efficiency but also improves system performance and adaptability. This section will examine the architecture of Microsoft Hyper-V, focusing on its fundamental components, functionalities, and design principles.

There are two primary categories of hypervisors: type 1 and type 2. A type 1 hypervisor operates as a streamlined operating system, functioning directly on the host's hardware. Conversely, a type 2 hypervisor acts as a software layer on an existing operating system. Type 1 hypervisors are considered more secure due to their separation from potentially vulnerable operating systems, and they generally exhibit superior performance compared to type 2 hypervisors. This makes type 1 hypervisors particularly well-suited for the demands of datacenter computing. While type 1 hypervisors run directly on the hardware, type 2 hypervisors operate atop the host device's operating system. Although type 2 hypervisors can support the installation of different operating systems, they experience higher latency than type 1 hypervisors. This increased latency arises from the necessity for communication to traverse the additional layer of the operating system. Type 2 hypervisors are often referred to as client hypervisors, as they are frequently utilized by developers for software testing and by end users. Hardware acceleration technology can enhance the speed of processing for both type 1 and type 2 hypervisors, enabling quicker generation and management of virtual resources.

Hyper-V features a type 1 hypervisor-based architecture. The hypervisor virtualizes processors and memory. It provides mechanisms for the virtualization stack in the root partition to manage child partitions and virtual machines (VMs) and expose services such as I/O (input/output) devices to the VMs.

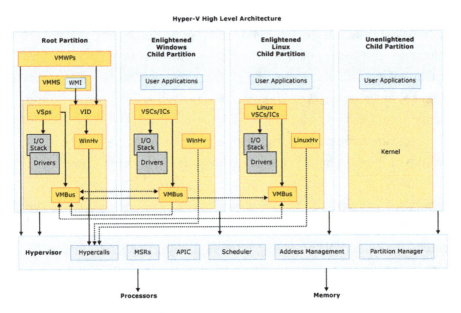

Figure 5-1. *High-level architecture of Hyper-V. Courtesy:* `https://learn.microsoft.com/en-us/virtualization/hyper-v-on-windows/reference/hyper-v-architecture`

5.1.8.1 Components of Hyper-V Architecture

Details of Hyper-V components are as follows.

Parent–Child Partition

The Hyper-V environment must include at least one host or parent partition. The virtualization stack operates within this parent partition, which has direct access to the underlying hardware. Subsequently, the parent partition creates guest virtual machines (child partitions), allowing for the installation of guest operating systems such as Windows and Linux. The guest VM does not have direct access to the physical processor nor manages the processor interrupts. Instead, the hypervisor manages all interrupts to the processor and redirects them to their corresponding guest VM.

VMBus

The Virtual Machine Bus (VMBus) is integral to the architecture of Microsoft Hyper-V, serving as a vital conduit for effective communication between virtual machines (VMs) and the host operating system. It functions as a high-speed interface that connects the virtualization stack with system services, thereby reducing latency and enhancing throughput. By leveraging the VMBus, Hyper-V optimizes performance in input/output operations and resource management, facilitating the efficient allocation of CPU, memory, and other resources among VMs. This architecture not only improves the overall performance of virtual environments but also streamlines the management of virtualized workloads, establishing it as a fundamental element of Hyper-V's framework. The VMBus operates as a channel-like inter-partition communication protocol, enabling the establishment of a communication link between the Hyper-V host and guest VM. It also plays a crucial role in machine enumeration on systems with multiple active virtualized partitions, effectively eliminating the need for any additional communication layers.

VSP–VSC

Virtual Service Client (VSC) and Virtual Service Provider (VSP) are the critical components of Hyper-V that facilitate easy and reliable communication between the Hyper-V Server and guest VMs. The VSPs always run in the parent partition (host) while the corresponding VSCs run in the child partition (guest VM). VSPs and their respective VSCs communicate with each other via VMBus. There can only be four VSPs (video, network, HID, and storage) running on an individual Hyper-V Server. However, multiple VSCs can run simultaneously on the same Hyper-V Server as a part of several guest VMs. VSPs are multithreading components that run as part of the VMMS.exe and can handle various VSC requests simultaneously.

VM Management Service (VMMS)

Virtual machine management service, also known as vmms.exe or VMMS, is the key module in Microsoft Windows OS that manages every aspect of Hyper-V Server virtualization. Virtual machine management service is a core Hyper-V component that runs under the system account, which has similar privileges to admin accounts. The Hyper-V virtual machine management service must be running, or the virtual machines will continue to run. But you won't be able to control, create, or delete virtual machines until the service runs. To summarize, VMMS is a core component of the Hyper-V platform and is responsible for

- Managing the state of all VMs in child partitions

- Creating a VM worker process (vmwp.exe) for each running VM

- Exposing a set of APIs based on Windows Management Instrumentation (WMI)

VM Worker Process

In the Microsoft Hyper-V architecture, the VM worker process plays a critical role in managing the execution of virtual machines (VMs). Each VM is associated with a dedicated VM worker process, responsible for handling tasks such as memory management, CPU scheduling, and I/O operations. This process interfaces with the Hyper-V hypervisor to ensure that resources are allocated efficiently and that VMs operate seamlessly while maintaining isolation from one another. By leveraging the VM worker process, Hyper-V can optimize performance and stability, enabling multiple VMs to run concurrently on a single physical host while minimizing the overhead associated with virtualization.

Hyper-V Guest Operating System Types

Hyper-V supports a number of different types of guest operating systems running in child partitions. Below are the three types:

> **Hyper-V Aware Windows operating systems:**
> Hyper-V aware Windows operating systems, often referred to as enlightened operating systems, have the ability to detect their execution on the Hyper-V hypervisor and modify their operations to improve performance. This is achieved through the use of hyper calls, which allow direct interaction with the hypervisor. Additionally, these operating systems are capable of supporting integration services, facilitating the operation of virtual service clients (VSCs) that communicate through the VMBus with virtual service providers (VSPs) situated in the root partition for device access.

Hyper-V Aware Non-Windows operating systems:
Operating systems that are not specifically tailored
for Windows Hyper-V can still leverage Integration
Services. Additionally, by utilizing third-party
virtual service clients (VSCs), they can access
devices via the virtual service providers (VSPs)
located in the root partition. These sophisticated
operating systems are also capable of modifying
their functions to improve performance and interact
directly with the hypervisor through hyper calls.

Non-Hyper-V Aware operating systems: These
operating systems do not recognize that they
are operating on a hypervisor and cannot utilize
the integration services. To accommodate these
systems, the Hyper-V hypervisor employs emulation
to facilitate access to device and CPU resources.
Although this method enables unmodified and
unenlightened operating systems to operate within
Hyper-V virtual machines, the overhead associated
with the emulation process can be considerable.

How Hyper-V Virtualization Works

Virtualization operates by separating or abstracting a computer's hardware
from the software that may run on it. This abstraction identifies the
physical resources of the computer – such as memory, processors, network
interfaces, and storage volumes – and creates logical representations. For
example, Hyper-V can transform a physical processor into a logical entity
referred to as a virtual CPU or vCPU.

Hyper-V is tasked with managing all the virtual resources it abstracts,
as well as overseeing the data exchanges between the physical resources
and their virtual equivalents. It utilizes these virtualized resources to create

logical representations of computers, known as virtual machines (VMs). Each VM is allocated memory, virtual processors, network adapters, storage, and other virtual components, including GPUs, all of which are managed by Hyper-V. When Hyper-V provisions a VM, the resulting logical instance operates independently of the underlying hardware and any other VMs created by Hyper-V. Consequently, a virtual machine remains unaware of other VMs or physical computers that may share the resources of the physical device.

This logical separation and meticulous resource management empower Hyper-V to create and oversee multiple VMs on a single physical machine. Each VM functions as a complete and operational computer, enabling organizations to derive multiple virtual servers from a single physical server. Upon the establishment of a VM by Hyper-V, a comprehensive suite of software must be installed, which includes drivers, an operating system, libraries, and appropriate enterprise applications. This capability allows organizations to run various operating systems to accommodate different workloads on the same physical computer.

Virtualization introduces a level of abstraction that grants virtual machines a degree of flexibility unattainable with traditional physical software and hardware setups. Each virtual machine operates within the physical memory of a computer, allowing for the storage of VMs as standard memory image files. These files can subsequently be utilized to replicate virtual machines on the same or different computers within the organization or to preserve the state of a VM at a specific moment in time. Additionally, it is possible to transfer a virtual machine from one virtualized environment to another. This process involves copying the selected VM from the memory of a source computer to the memory of a target computer, followed by the removal of the original VM from the source. Typically, this transfer can be executed without disrupting the operation of the VM or the experience of the user. The number of virtual machines that can be created is limited by the physical resources of the host computer and the computational demands of the enterprise applications running within those VMs. After a VM is established, its abstracted resources can be modified to enhance performance and facilitate the creation of additional VMs on the system.

5.1.9 Key Features of Hyper-V

In a time when organizations are persistently pursuing greater efficiency, scalability, and cost savings, virtualization has become a transformative technology. Microsoft Hyper-V stands out as one of the premier hypervisors available, offering a powerful platform for the creation, management, and operation of virtual machines. Whether you are a novice in the realm of virtualization or aiming to improve your current infrastructure, gaining insight into the essential features of Microsoft Hyper-V can enable you to fully leverage its capabilities. Let us put some more light on the key features provided by Hyper-V.

5.1.9.1 Virtual Machine Management

At the heart of Microsoft Hyper-V is its powerful virtual machine management capability. Hyper-V allows you to create and manage virtual machines (VMs) with ease. Through the Hyper-V Manager or Windows Admin Center, users can set up new VMs, configure their settings, allocate resources, and manage their lifecycle. The intuitive interface simplifies the process, making it accessible even to those with minimal IT experience.

5.1.9.2 Dynamic Memory

Dynamic memory in Microsoft Hyper-V is a feature that optimizes the allocation and use of memory for virtual machines (VMs) by allowing them to adjust their memory requirements dynamically based on workload demands. This capability enables Hyper-V to efficiently allocate physical memory between multiple VMs, ensuring that resources are used effectively and enhancing overall performance. Administrators can set minimum and maximum memory limits for each VM, and Hyper-V will automatically adjust the memory allocation as needed, allowing for better resource utilization and improved responsiveness in environments with fluctuating workloads. This dynamic adjustment not only helps in maximizing the physical memory on the host but also supports more flexible and efficient management of virtualized workloads.

5.1.9.3 Live Migration

Live migration in Microsoft Hyper-V is a feature that enables the seamless transfer of virtual machines (VMs) between physical hosts without service interruption. This technology allows administrators to maintain high availability and optimize resource utilization by redistributing workloads in real time. During the migration process, the VM's running state, including its memory and network connections, is transferred in small chunks while continuing to operate, making it transparent to end users. Live migration is particularly beneficial in scenarios such as load balancing, hardware maintenance, and disaster recovery, enhancing overall system reliability and performance within a virtualized environment.

5.1.9.4 Virtual Switches

Hyper-V virtual switches play a crucial role in facilitating network connectivity for virtual machines (VMs). These virtual switches simulate physical network switches, enabling VMs to communicate with each other and with external networks. Hyper-V supports three types of virtual switches: external, which connects VMs to a physical network; internal, which allows communication between VMs and the host server; and private, which restricts VM communication to each other only. This flexible architecture enhances the management of network traffic, security, and isolation, allowing users to create diverse network configurations according to their needs.

5.1.9.5 Hyper-V Replica

Hyper-V Replica is a disaster recovery feature in Microsoft Hyper-V that enables the replication of virtual machines (VMs) from one Hyper-V host to another, providing business continuity in the event of an outage. By periodically copying VM data to a secondary location, Hyper-V Replica ensures that an up-to-date version of the virtual machine is available for failover. This asynchronous replication occurs over standard network connections and can be configured for different recovery point objectives (RPOs), allowing organizations to tailor the level of data protection to their specific needs. In addition, Hyper-V Replica supports the capability to perform non-disruptive test failovers, enabling IT administrators to validate their disaster recovery plans without impacting production workloads.

5.1.9.6 Nested Virtualization

Nested virtualization is a feature in Microsoft Hyper-V that allows you to run Hyper-V virtual machines (VMs) inside other Hyper-V VMs. This capability is particularly valuable for development and testing environments, as it enables users to create, manage, and test virtualization scenarios without requiring dedicated physical hardware. With nested virtualization, developers can simulate hypervisor workloads, perform testing of virtualized applications, and evaluate cloud services and deployment strategies all within a secured and isolated VM environment. This feature enhances flexibility and efficiency in managing complex virtualization setups, making it easier for IT professionals to experiment and prototype solutions in a safe manner.

5.1.9.7 Integration Services

Integration services in Microsoft Hyper-V are a suite of tools that enhance the performance and functionality of virtual machines (VMs) running on the Hyper-V platform. These services facilitate seamless interaction between the host and guest operating systems, ensuring improved management and operational efficiency. Key features include time synchronization, data exchange, heartbeat monitoring, and backup services, which help maintain system integrity and optimize resource usage. Additionally, Integration Services enable smooth handling of network adapters and storage, providing a rich virtual machine experience that mimics physical machines. By incorporating these services, users can achieve better performance and manageability within their virtualized environments.

5.1.9.8 Fault Tolerance and High Availability

Fault tolerance and high availability in Microsoft Hyper-V are essential features designed to ensure continuous operation and minimal downtime for virtualized environments. Hyper-V's high availability utilizes clustering technology, allowing virtual machines (VMs) to automatically failover to other nodes in a cluster in case of hardware or software failures, thus maintaining service continuity. Coupled with features like Live Migration, which enables the seamless movement of VMs between hosts without interruption, these capabilities help organizations achieve robust resilience against system failures. Additionally, Hyper-V's integration with Windows Server provides automated backup and recovery options, further enhancing the reliability of virtual environments and ensuring that critical workloads remain accessible even during unexpected disruptions.

5.1.9.9 Virtual Hard Disk Options

Microsoft Hyper-V offers several options for managing virtual hard disks (VHDs), crucial for effective virtual machine configuration and storage optimization. Hyper-V supports both VHD and VHDX formats, with VHDX providing benefits like larger storage capacity, improved performance, and data corruption protection through logging. Users can choose between fixed-size disks, which allocate all space upfront, and dynamically expanding disks, which grow as needed up to a set limit. Additionally, Hyper-V allows the use of differencing disks, enabling snapshots for backup and testing purposes without duplicating entire VHD files. These options empower IT administrators to tailor storage solutions that enhance performance and resource efficiency in virtual environments.

5.1.9.10 Enhanced Security Features

Microsoft Hyper-V offers enhanced security features that significantly bolster the protection of virtualized environments. One of the key components is Secure Boot, which ensures that only trusted operating system loaders can run during the boot process, preventing unauthorized access. Additionally, Shielded VMs provide further security by encrypting virtual machine files, isolating them from potentially hostile administrators. This is complemented by the use of virtual TPM (trusted platform module), which enables secure storage of cryptographic keys and enhances data protection. With features like just-in-time (JIT) access and just-enough-administration (JEA), Hyper-V also minimizes the risk of privilege escalation, creating a more secure and resilient virtualization landscape. Overall, these features work in synergy to safeguard workloads against evolving cyber threats while maintaining compliance with security standards.

Microsoft Hyper-V transcends the role of a mere virtualization platform; it serves as a holistic solution that enables organizations to enhance their IT infrastructures. With features such as dynamic memory management, seamless live migration, advanced networking, and strong security protocols, Hyper-V is tailored to address the requirements of contemporary enterprises. As you delve into the functionalities of Hyper-V, you will discover that its extensive array of features not only boosts performance but also aligns with your strategic objectives of scalability, efficiency, and resilience in a rapidly changing technological environment. Regardless of whether you operate a small business or oversee a large corporation, leveraging the capabilities of Microsoft Hyper-V can facilitate operational excellence and yield cost efficiencies, thereby ensuring your organization remains competitive in the fast-evolving digital landscape.

5.2 Comparison of Hyper-V and VMware

As virtualization technology has become essential for businesses seeking efficiency, flexibility, and cost savings, let's discuss and compare two dominant players in the market: Microsoft with Hyper-V and VMware with its suite of virtualization products. While both serve similar purposes, they differ significantly in architecture, performance, usability, features, scalability, and cost. This section provides an in-depth comparison of Hyper-V and VMware across these critical dimensions.

Table 5-1. *Comparison of VMware and Hyper-V*

Category	Vmware	Hyper-V
Architecture	Type-1 Hypervisor	Type-1 Hypervisor
	Light weight kernel that runs only from memory after boot	Feature add-on to windows server and windows desktop version. Requires installation of Windows OS to function
Performance	Essentially Equal performance to Hyper-V	Essentially Equal performance to Vmware
	ESXi 8 supports a maximum of 896 logical CPUs and upto 24 TB of RAM per host	Hyper-V supports up to 512 logical CPUs and up to 48 TB of RAM (windows server 2022)
Usability	Built-in HTML5 web-based host management GUI	Host management GUI is built-in, provided by Microsoft Management console
	ESXi's host management GUI is focused on host and virtual machine management, and you can open in browser.	Hyper-V host administration must be done via windows-based system (called Hyper-V manager) or via powershell command line as well.
Features	Distributed Resource Scheduling (DRS)	Workload Balancing
	High Availability (HA), Fault tolerance	High - Availability, Host clutering
	Live Migration of VM's (vMotion)	Live Migration of VMs
	Live Storage vMotion	Live Storage Migration
	Guest VM's requires Additionl licensing	*With windows datacenter host running Hyper-V, Windows guest operationg systems runs "Free"*
Scalability	ESXi capable of managing 1024 VMs and with vCenter it can support 96 hosts in a cluster, with Vmware vSAN it is 64 nodes in a cluster.	Hyper-V supports over one thousan VM's per host. Hyper-V cluster manage up to 64 nodes per cluter with up to 8,000 total VMs.
Cost	Recent anouncement with Broadcom removed access to free ESXi hypervisor	Hyper-V is a free add-on to Microsoft windows server
	Vmware Cloud Foundation (VCF): MSRP $350/core == $5,600/16 core. Guest VM licenses are extra	With Windows Datacenter server operating system allows unlimited Windows VM's at no extra license cost. Windows Datacenter MSRP cost is $6155 for 16 cores
		With Windows Standard server operating system allows two Windows VM's at no extra license cost. Windows Server MSRP cost is $1,069 for 16 cores

5.2.1 Architecture

Hyper-V architecture: Hyper-V, a virtualization technology developed by Microsoft, is deeply integrated into the Windows Server ecosystem, providing a robust architecture that leverages existing Windows features and services. Hyper-V employs a type 1 hypervisor, integrated into Windows Server, and utilizes a microkernel architecture. This design allows for a lightweight hypervisor that minimizes resource overhead. Hyper-V runs at the kernel level, which benefits from Windows' established security features. It supports a range of guest operating systems, including various versions of Windows and Linux.

VMware architecture: VMware also employs a type 1 hypervisor known as ESXi that operates directly on the host hardware without the need for a base operating system. This architecture enhances performance and stability, enabling the hypervisor to deliver near-native functionality. VMware vSphere, which includes ESXi and other tools, can manage large environments effectively and integrates tightly with its comprehensive suite of products.

5.2.2 Performance

Hyper-V performance: Hyper-V is known to provide competitive performance for typical workloads. It is used at a large-scale enterprise organization across the globe. Microsoft also continually improves performance with each version, particularly concerning network throughput and storage integration.

VMware performance: VMware is also known for its performance, particularly with resource-intensive applications and large-scale virtualization scenarios. Features like vMotion, distributed resource scheduler (DRS), and more advanced memory management contribute to its performance, allowing efficient load balancing and resource allocation.

5.2.3 Usability

Hyper-V usability: Hyper-V management integrates seamlessly with Windows Server Manager and PowerShell, offering familiar tools for Windows administrators. Hyper-V has made significant strides in this area with the Windows Admin Center providing the capability to manage Hyper-V clusters. SCVMM (system center virtual machine manager) is another tool from Microsoft that offers great manageability of multiple Hyper-V clusters and large-scale setups.

VMware usability: VMware's vCenter Server provides a centralized management interface that is robust yet intuitive. While the learning curve can be steeper for newcomers, experienced users find it highly functional for conducting detailed analysis and managing complex virtual environments. VMware vCenter is also a great tool that simplifies the manageability of multiple ESXi clusters in a large-scale setup.

5.2.4 Features

Hyper-V features: Hyper-V offers a range of features; some notable ones include

- **Nested virtualization:** Allows running Hyper-V within a virtual machine.

- **Live migration:** Moves VMs between hosts with minimal downtime.

- **Virtual switches and storage space direct (SSD):** Customizable networking options for security and performance. SSD is the software-defined storage solution.

- **Hyper-V checkpoints:** For faster backups. Also, the technology offers application-consistent backups as well.

- **Integration with Azure:** Facilitates hybrid cloud environments.

VMware features: VMware is rich in features, several of which include:

- **Nested virtualization:** Allows running ESXi within a virtual machine.

- **VMware tools:** Enhances management and performance for VMs.

- **vMotion and storage vMotion:** Allows live migration of VMs without downtime.

- **Snapshot and cloning:** Facilitates rapid backups and deployment.

- **NSX and vSAN:** Advanced networking and storage integration for cloud environments.

5.2.5 Scalability

Hyper-V scalability: Hyper-V is highly scalable, supporting hundreds of VMs on a single host. It provides features like dynamic memory to optimize VM sizes based on demand. Microsoft's focus on hybrid cloud capabilities boosts its scalability when integrated with Azure.

VMware scalability: VMware is also supporting thousands of VMs across clusters. vSphere is designed to manage vast data centers and is mainly for on-premises workloads.

5.2.6 Cost

Hyper-V cost: Hyper-V is often seen as a cost-effective solution, especially for organizations already using Windows Server. It is typically included with Windows Server licenses, minimizing the expenditure for those environments. Microsoft's licensing model tends to be more straightforward, enhancing affordability for enterprises.

VMware cost: VMware solutions typically come with higher licensing fees compared to Hyper-V. While VMware's feature set may justify the cost for enterprise-level organizations, smaller firms may find it a significant investment. Additionally, VMware's various licensing options can complicate budgeting for businesses that need flexibility.

Both Hyper-V and VMware are powerful virtualization platforms, each with distinct strengths and weaknesses. Hyper-V shines in affordability and integration with Microsoft services, making it an attractive choice for businesses invested in the Windows ecosystem. Ultimately, the choice between Hyper-V and VMware will depend on the specific needs of the organization, including existing infrastructure, budget constraints, and performance requirements. Understanding the differences and aligning them with business goals can help decision-makers select the right virtualization technology to meet their objectives effectively.

5.3 Deconstructing Microsoft Azure Stack Hyper-converged Infrastructure (HCI)

Azure Stack HCI (or Azure Local) represents a hybrid cloud solution that merges the capabilities of Microsoft Azure's cloud services with local datacenter assets. At its essence, HCI streamlines the deployment and oversight of virtualized workloads by integrating computing, storage, and networking into a unified framework. This integration results in simplified management, a smaller physical footprint, and improved performance. The core technology of Azure Stack HCI is based on Windows Server and Hyper-V, which allows organizations to operate virtual machines (VMs) in conjunction with native Azure services. Moreover, Azure Stack HCI provides seamless connectivity with the Azure ecosystem, promoting hybrid operations. This feature enables enterprises to utilize cloud services such as Azure Backup, Azure Site Recovery, and various monitoring tools directly from their on-premises setup. The capacity to access native Azure services while retaining control over local resources establishes a flexible platform that accommodates diverse operational requirements, ranging from edge computing to data analytics. Security remains a critical focus in HCI environments, and with Azure Stack, organizations gain access to robust security features, including built-in encryption and advanced threat protection, thereby ensuring data integrity and adherence to industry regulations. Moreover, scalability and flexibility are inherent in Azure Stack HCI, enabling organizations to adapt to changing demands. Users can easily scale their infrastructure by adding nodes, adjusting resources dynamically, and optimizing workloads based on performance metrics. Additionally, Azure Stack HCI supports various deployment scenarios, including virtual desktop infrastructure (VDI) and containerized applications, making it suitable for diverse organizational needs. As companies increasingly adopt hybrid cloud strategies, understanding the nuances of Azure Stack HCI is crucial for maximizing its benefits and ensuring a robust, integrated IT framework that enhances operational agility and drives innovation.

5.3.1 Hybrid Cloud and Microsoft Azure Stack HCI

The trajectory of enterprise IT is clearly moving toward hybrid cloud solutions, as organizations increasingly acknowledge the advantages of integrating on-premises infrastructure with both public and private cloud services. This hybrid model enables businesses to enhance their resource utilization by leveraging the scalability and adaptability offered by cloud environments while maintaining oversight of sensitive data and essential applications. In their pursuit of enhanced agility and innovation, companies can swiftly respond to evolving market demands, rapidly launch new services, and propel comprehensive digital transformation efforts within hybrid cloud settings. As technological advancements continue to unfold, the hybrid cloud is poised to become a fundamental element of contemporary IT architecture. Industry leaders are already channeling investments into platforms that facilitate multi-cloud strategies, allowing businesses to select the optimal combination of services and tools that align with their goals. In this ever-changing landscape, collaborations between cloud service providers and enterprises will be vital, promoting the integration of emerging technologies such as artificial intelligence, machine learning, and the Internet of Things (IoT). Ultimately, the hybrid cloud not only grants organizations the flexibility to innovate but also positions them for enduring growth and success in a progressively competitive digital marketplace.

Industry research underscores the increasing implementation of hybrid cloud solutions across diverse sectors, motivated by the rising demand for data-driven decision-making and uninterrupted business operations. Studies reveal that organizations utilizing hybrid cloud environments are more adept at leveraging cutting-edge technologies such as artificial intelligence and machine learning, as they can adjust resources in accordance with project needs without excessive capital investment. Furthermore, the hybrid cloud model enhances disaster recovery initiatives, enabling businesses to sustain operations during unexpected disruptions by utilizing cloud-based backup resources.

Security continues to be a critical area of focus in hybrid cloud research, with numerous studies examining the strategies organizations adopt to safeguard their data across various environments. The shared responsibility model inherent in hybrid cloud configurations requires a comprehensive security framework that includes identity management, data encryption, and adherence to regulations like GDPR or HIPAA. Research shows that organizations that implement hybrid cloud solutions frequently allocate resources to advanced security tools and practices to reduce the risks associated with data breaches and unauthorized access.

Recent studies extensively examine the financial implications of adopting hybrid cloud solutions, indicating that companies can realize considerable cost savings through more efficient resource management. Hybrid clouds enable organizations to pay solely for the resources they consume, thereby removing the necessity for substantial initial investments in IT infrastructure. Furthermore, findings indicate that the adoption of a hybrid cloud strategy frequently results in lower operational expenses and enhanced return on investment (ROI), rendering it an attractive option for organizations aiming to secure their future operations and advance their digital transformation efforts.

Microsoft Azure Stack HCI is a significant player in the field, offering a solution that allows enterprises to extend Azure services into their on-premises datacenters. This platform combines the advantages of hyper-converged infrastructure (HCI) with the capabilities of Azure's cloud, creating a cohesive environment that supports seamless application development, deployment, and management across both local and cloud settings. Microsoft Azure Stack HCI leverages the strengths of software-defined storage, networking, and virtualization to create a highly efficient and performance-oriented infrastructure. Organizations can run virtual machines, containers, and microservices while maintaining complete control over their data and applications. This flexibility is particularly valuable for businesses with sensitive data that must remain on-site due to regulatory constraints. Additionally, Azure Stack HCI enables enterprises to take advantage of Azure's cloud services for backup, disaster recovery, and scalability, allowing them to respond quickly to changing demands without incurring the latency that can arise from purely cloud-based solutions.

The integration of Azure Stack HCI with the wider Azure ecosystem streamlines management and orchestration processes. IT teams can utilize familiar Azure tools and interfaces to oversee, deploy, and manage resources, thereby reducing the steep learning curve typically linked to new technology adoption. Additionally, organizations can take advantage of Azure's advanced analytics and artificial intelligence features to extract insights from their data, whether stored on-premises or in the cloud. As hybrid cloud solutions continue to advance, Microsoft Azure Stack HCI emerges as a robust enabler, empowering businesses to cultivate a more agile and resilient IT infrastructure while optimizing their investments in both on-premises and cloud resources.

5.3.2 History of Microsoft Azure Stack HCI

Microsoft Azure Stack HCI, also known as hyper-converged infrastructure, signifies a notable advancement in the deployment and management of IT infrastructure by organizations. Introduced in 2019, Azure Stack HCI was developed in response to Microsoft's recognition that businesses were increasingly seeking to merge cloud solutions with on-premises functionalities. This hybrid model was motivated by the demand for flexibility and scalability in a landscape where data and applications are dispersed across multiple environments. By integrating virtualization, software-defined storage, and networking into a unified solution, Azure Stack HCI allows organizations to retain their essential workloads on-premises while effortlessly connecting to the Azure cloud.

The origins of Azure Stack HCI can be linked to the broader transition toward cloud computing that commenced in the late 2000s and gained momentum in the subsequent decade. As organizations began migrating to the cloud, they faced challenges concerning latency, compliance, and data control. In response to these issues, Microsoft created the Azure Stack portfolio, enabling businesses to extend Azure services into their local datacenters. Azure Stack HCI was specifically designed as the hypervisor-based solution within this portfolio, addressing the needs of organizations requiring improved performance and reliability for their virtualized workloads.

Over the years, Azure Stack HCI has undergone significant evolution driven by ongoing innovation, user input, and the latest technological developments. It is founded on Windows Server and employs a hyper-converged architecture that seamlessly combines compute, storage, and networking into a unified platform. Furthermore, Microsoft has highlighted the significance of a strong ecosystem by collaborating with hardware vendors, thereby offering customers a range of certified solutions customized to meet their unique requirements. Through regular updates and the rollout of new features, Azure Stack HCI has solidified its position as a fundamental element of Microsoft's hybrid cloud strategy, empowering organizations to manage their IT operations more efficiently across both on-premises and cloud settings.

5.3.2.1 Azure Stack HCI 20H2

Released in October 2020, the 20H2 update represented a pivotal shift for Azure Stack HCI, transitioning to a subscription-based model. This iteration aimed to streamline the deployment process and enhance integration with Azure services. A notable feature of Azure Stack HCI 20H2 is its robust integration with Azure, enabling organizations to utilize cloud functionalities such as Azure Monitor, Azure Security Center, and Azure Backup, all while retaining control over their on-premises data. This hybrid architecture allows businesses to remain agile, scaling resources as necessary to maximize their IT investments. Moreover, Azure Stack HCI 20H2 prioritizes enhanced performance and manageability. It supports the latest hardware advancements, including high-speed NVMe storage and sophisticated networking options, which contribute to improved throughput and reduced latency – essential for data-intensive applications. The graphical user interface (GUI) has also been refined to simplify deployment and management processes, facilitating easier configuration and maintenance for IT teams. Additionally, it includes support for Kubernetes, which aids in the consistent deployment of modern applications and services across both on-premises and cloud environments, thereby promoting innovation and decreasing time to market.

Security continues to be a paramount concern in Azure Stack HCI 20H2, which integrates capabilities like Windows Defender Advanced Threat Protection (ATP) to defend against new and evolving threats. Organizations are able to establish strong security protocols while meeting compliance requirements, thereby ensuring the protection of their data, whether it is stored in the cloud or on local servers. By connecting on-premises infrastructures with Azure, Azure Stack HCI 20H2 enables enterprises to pursue digital transformation, respond to shifting market needs, and improve their operational effectiveness in a highly competitive environment.

Key Features

- **Integration with Azure Monitor:** Organizations can monitor their Azure Stack HCI environment through Azure Monitor, gaining insights into health performance and storage.

- **Azure policy for consistency:** Enforce policies on Azure Stack HCI clusters to maintain security and compliance.

- **Nested virtualization:** Support for running additional Hyper-V virtual machines within VMs, enabling new use cases such as testing and development.

5.3.2.2 Azure Stack HCI 21H2

Launched in September 2021, Azure Stack HCI 21H2 represents a major enhancement to Microsoft's hyper-converged infrastructure offering, aimed at providing enterprises with a scalable and adaptable platform for managing virtualized workloads. This version introduces improved performance, better integration with Azure services, and updated functionalities tailored for hybrid cloud environments. A notable feature of Azure Stack HCI 21H2 is its compatibility with Azure Arc, enabling users to manage both on-premises and Azure resources in a unified manner. This integration streamlines management processes and bolsters security by ensuring consistent governance across hybrid settings. The 21H2 release also enhances hardware compatibility, accommodating the latest generations of servers and storage technologies, thus allowing organizations to take advantage of cutting-edge infrastructure innovations. Additionally, it features improved virtualization capabilities through Windows Server 2022, which facilitates enhanced resource allocation, fault tolerance, and disaster recovery options. The inclusion of Kubernetes support further enables organizations to efficiently run containerized applications, aligning their operations with contemporary DevOps methodologies and cloud-native frameworks. Beyond these technical improvements, Azure Stack HCI 21H2 prioritizes user experience with a refined management interface accessible via Windows Admin Center and Azure Portal. This intuitive design aids IT professionals in effectively monitoring and managing their hyper-converged infrastructure. Continuous updates and support from Microsoft not only expand the feature set but also provide organizations with assurances of ongoing security and performance enhancements. As businesses increasingly embrace hybrid cloud strategies, Azure Stack HCI 21H2 emerges as an essential solution for organizations aiming to optimize their infrastructure while retaining flexibility and control.

Key Features

- **Improved security with Windows Server:** Enhanced security features, including secured-core configuration and advanced threat protection.

- **Azure arc integration:** Management of Azure Stack HCI resources through Azure Arc, enabling a unified management experience across multiple environments.

- **Storage migration service enhancements:** Simplified migration of file servers to Azure Stack HCI with additional capabilities to reduce downtime and improve reliability.

5.3.2.3 Azure Stack HCI 22H2

The 22H2 update, launched in September 2022, significantly enhanced the capabilities of the platform, particularly in the realms of edge computing and management functionalities. Azure Stack HCI 22H2 represents a major advancement in Microsoft's hyper-converged infrastructure (HCI) offering, aimed at improving on-premises virtualization and its integration with Azure services. This iteration builds upon the solid foundation established by its predecessor, introducing superior performance, enhanced security measures, and a more profound integration with Azure, making it an optimal choice for organizations seeking to modernize their infrastructure. With its capacity to efficiently run virtual machines and containerized applications, Azure Stack HCI 22H2 accommodates a variety of workloads, enabling organizations to optimize resource utilization and enhance operational efficiency.

A notable feature of Azure Stack HCI 22H2 is its refined management experience via the Azure portal, which enables administrators to manage their hybrid environments with ease. The update offers improved monitoring, diagnostics, and automated updates, thereby minimizing administrative burdens and allowing IT teams to concentrate on strategic projects rather than routine maintenance tasks. Additionally, the introduction of enhanced scaling options and support for various hardware configurations ensures that organizations can tailor their deployments to fulfill specific performance and compliance needs. Moreover, Azure Stack HCI 22H2 emphasizes Microsoft's dedication to security by incorporating advanced security features, such as Azure Defender, which provides threat protection for workloads operating on the HCI solution. Organizations can take advantage of these built-in security measures without sacrificing operational agility. As cloud-native applications gain prominence and the demand for hybrid solutions increases, this update positions Azure Stack HCI as a compelling choice for enterprises looking to harness the benefits of the cloud while retaining on-premises control and compliance. Overall, Azure Stack HCI 22H2 represents a significant step forward in the evolution of hybrid infrastructure solutions.

Key Features:

- **Enhanced support for Kubernetes:** Improved Kubernetes deployment and management capabilities, including Azure Kubernetes Service (AKS) on Azure Stack HCI.

- **Storage resiliency:** Introduced features that improve resilience and performance in storage solutions, including support for larger volumes.

- **Data performance enhancements:** Improved IOPS performance for workloads such as SQL Server and other data-intensive applications.

5.3.2.4 Azure Stack HCI 23H2

Microsoft continues to enhance its Azure Stack HCI platform. The release of Azure Stack HCI 23H2 introduces a suite of new features that aim to improve performance, expand functionality, and streamline management for IT teams. With 23H2, Microsoft brought considerable additions to the product; let's discuss them in brief below.

Enhanced Performance and Scalability

The 23H2 release of Azure Stack HCI comes with significant improvements in performance and scalability, which are critical for businesses seeking to optimize their IT infrastructure. Key enhancements include

> **Improved storage performance**: With advanced storage options, including support for NVMe over Fabrics (NVMe-oF), Azure Stack HCI 23H2 significantly boosts storage performance. This translates to higher throughput and lower latency, which is particularly advantageous for data-intensive applications and workloads.

> **Scalability improvements**: Organizations can now scale their deployments more seamlessly, allowing them to cater to changing workloads and business needs. The latest version supports up to 16 nodes in a single cluster, empowering businesses to expand their resources without disrupting operations.

New Management Features

Simplified management is a critical aspect of any hybrid cloud solution, and Azure Stack HCI 23H2 delivers with several key features:

> **Enhanced Windows Admin Center integration**: The integration with Windows Admin Center has been refined, providing IT administrators with a more intuitive and user-friendly interface for managing and monitoring HCI clusters. This includes comprehensive dashboards that highlight system health, performance metrics, and alerts.

Automated deployment and configuration:
The new version includes improved automated deployment and configuration tools. With the Azure Stack HCI installation wizard, IT teams can rapidly install and configure clusters, reducing the time and effort required to deploy new solutions.

Centralized management with Azure Arc: Azure Stack HCI 23H2 is designed to work seamlessly with Azure Arc, allowing organizations to manage on-premises and cloud resources from a unified platform. This centralized management approach improves visibility and control, enabling better governance and compliance.

Advanced Networking Capabilities

Networking is pivotal in hybrid cloud environments, and Azure Stack HCI 23H2 brings several innovations to improve connectivity:

Enhanced software-defined networking (SDN):
The updated version features advanced SDN capabilities that simplify the creation, management, and monitoring of virtual networks. This improves network security and performance while allowing for more efficient resource allocation.

Improved load balancing: With enhanced load balancing features, Azure Stack HCI 23H2 ensures optimal resource distribution across applications and services. This results in improved application performance and user experience by minimizing latency and maximizing uptime.

Support for Kubernetes and Cloud-Native Applications

As the demand for cloud-native applications and microservices architecture increases, Azure Stack HCI 23H2 provides native support for Kubernetes, making it easier for developers to deploy and manage containerized applications:

Integrated Kubernetes support: The new version offers built-in Kubernetes integration, enabling organizations to run containerized applications alongside virtual machines within the same infrastructure. This flexibility fosters a DevOps culture by streamlining development and operations.

Enhanced developer experiences: The integration of developer tools and environments enhances the experience for teams engaging in agile development practices. This includes improved support for CI/CD pipelines and seamless deployment of applications using Azure DevOps.

Security and Compliance Enhancements

Security is a top concern for organizations embracing hybrid cloud solutions. Azure Stack HCI 23H2 includes several cybersecurity features designed to protect data and ensure compliance with regulations:

Advanced threat protection: Leveraging Microsoft's security intelligence, Azure Stack HCI offers enhanced threat detection and response capabilities. This allows organizations to identify and remediate vulnerabilities more effectively, safeguarding their systems from evolving threats.

Compliance management: The new Compliance Manager provides tools for monitoring and maintaining compliance with various industry standards and regulations. This is essential for organizations operating in regulated environments, helping them adhere to best practices and mitigate risks.

The launch of Azure Stack HCI 23H2 represents a pivotal advancement in the development of hybrid cloud solutions. Featuring substantial performance improvements, enhanced management functionalities, superior networking options, integrated Kubernetes support, and strengthened security protocols, this latest iteration enables organizations to fully realize their hybrid cloud strategies. As enterprises strive to manage the intricacies of contemporary IT landscapes, Azure Stack HCI 23H2 emerges as a formidable platform that facilitates innovation, scalability, and success in the digital era.

5.3.3 Architecture of Microsoft Azure Stack HCI

Azure Stack HCI represents a hyper-converged infrastructure (HCI) solution designed to support both Windows and Linux virtual machines or containerized workloads, along with their associated storage. This hybrid solution facilitates a connection between on-premises systems and Azure, enabling access to cloud-based services, monitoring, and management capabilities. An Azure Stack HCI setup typically comprises a single server or a cluster of servers operating on the Azure Stack HCI operating system, all linked to Azure. Users can utilize the Azure portal to oversee and manage individual Azure Stack HCI systems, as well as to review all deployments of Azure Stack HCI. Additionally, management can be conducted using existing tools such as Windows Admin Center and PowerShell. To obtain the necessary servers for running Azure Stack

HCI, one can either acquire Azure Stack HCI integrated systems from a Microsoft hardware partner, which come with the operating system pre-installed, or purchase validated nodes and install the operating system independently.

Azure Stack HCI is built on proven technologies, including Hyper-V, Storage Spaces Direct, and core Azure management service. Each Azure Stack HCI system consists of between 1 and 16 physical servers. All servers share common configurations and resources by leveraging the Windows Server Failover Clustering feature. Azure Stack HCI combines the following:

- Validated hardware from a hardware partner

- Azure Stack HCI operating system

- Hyper-V-based compute resources

- Storage Spaces Direct-based virtualized storage

- Windows and Linux virtual machines as Arc-enabled servers

- Azure Virtual Desktop

- Azure Kubernetes Service (AKS) enabled by Azure Arc

- Azure services, including monitoring, backup, site recovery, and more

- Azure portal, Azure Resource Manager and Bicep templates, and Azure CLI and tools

Figure 5-2. *Azure Stack HCI high-level architecture diagram.*
Courtesy: Microsoft Online documentation on Azure Stack HCI.

5.3.3.1 Key Components of Azure Stack HCI

At its core, Azure Stack HCI is built upon a hyper-converged architecture. This means it combines compute, storage, and networking into a single software-defined solution. It utilizes

- **Windows server**: Azure Stack HCI is powered by Windows Server Datacenter, specifically incorporating features from Windows Server 2022 and later versions.

- **Storage Spaces Direct (S2D)**: An integral component that provides software-defined storage. It allows the creation of highly available and scalable storage solutions using local storage across clustered nodes, eliminating the need for separate storage appliances.

- **Virtualization**: Leveraging Hyper-V technology, Azure Stack HCI supports a variety of workloads, from traditional virtual machines to modern applications running in containers.

5.3.3.2 Management Tools

Azure Stack HCI provides powerful management tools to streamline operations:

- **Windows Admin Center (WAC)**: A web-based interface that gives administrators real-time insights, management capabilities, and a unified view of their Azure Stack HCI environment. WAC allows users to manage virtual machines, storage, and overall cluster health with ease.

- **Azure portal integration**: Azure Stack HCI integrates natively with the Azure Portal, providing access to a broad array of Azure services. Users can monitor, manage resources, and implement consistent policy management across their Azure and on-premises environments.

5.3.3.3 Networking

Networking components in Azure Stack HCI involve

- **Software-defined networking (SDN)**: Azure Stack HCI integrates with Windows Server's networking capabilities, allowing organizations to create virtual networks and domains. It supports features like Virtual Network Gateways, Load Balancers, and Network Security Groups (NSGs) for enhanced security and traffic management.

- **Azure networking services**: Through Azure integration, users can leverage Azure services such as Azure VPN Gateway for site-to-site connectivity and Azure ExpressRoute for private network connections.

5.3.3.4 Security Features

Security is paramount for any cloud or hybrid solution. Azure Stack HCI includes:

- **Azure Defender**: An integrated threat protection solution that helps protect workloads across both Azure and on-premises environments.

- **Azure Security Center**: Offers a unified security management system offering advanced threat protection across hybrid workloads in the cloud, as well as on-premises systems.

5.3.3.5 Integrations with Azure Services

One of the significant advantages of Azure Stack HCI is its native cloud integration capabilities:

- **Azure Backup:** Organizations can utilize Azure Backup to protect and recover on-premises workloads, ensuring data availability during disasters or unexpected outages.

- **Azure Site Recovery:** This service allows businesses to automate the replication and recovery of virtual machines between on-premises and Azure, ensuring business continuity in case of disasters.

- **Azure Monitor and Log Analytics:** Azure Stack HCI integrates with Azure Monitor to provide detailed performance metrics and operational logs, enabling administrators to proactively manage resources.

- **Azure Kubernetes Service (AKS):** For those looking to implement containerized applications, Azure Stack HCI supports deploying Azure Kubernetes Service, allowing organizations to run and manage Kubernetes clusters in their on-premises environments.

The architecture of Microsoft Azure Stack HCI represents a strategic evolution in how businesses are approaching their IT infrastructure requirements. By enabling organizations to take advantage of a hyper-converged model that is tightly integrated with Azure's cloud services, Microsoft provides a platform that fosters agility and efficiency while minimizing complexity. As businesses increasingly look to adopt hybrid cloud solutions to meet their evolving needs, Azure Stack HCI stands out for its versatility, ease of management, and robust feature set that makes it a cornerstone of modern IT strategies. Whether it's enhancing on-premises resources or bridging the gap to the cloud, Azure Stack HCI delivers the tools and integrations necessary to propel organizations into the future of computing.

5.4 Comparison of Microsoft Azure Stack HCI and VMware Cloud Foundation

Microsoft Azure Stack HCI and VMware Cloud Foundation stand out as two prominent options, each designed to address distinct requirements, functionalities, and implementation approaches. This section will examine the fundamental differences, benefits, and applicable scenarios of these two cloud solutions, offering valuable insights to assist organizations in making well-informed choices regarding their cloud infrastructure.

Figure 5-3 provides a comprehensive feature-wise comparison between VMware Cloud Foundation offerings and Microsoft Azure Stack HCI offerings.

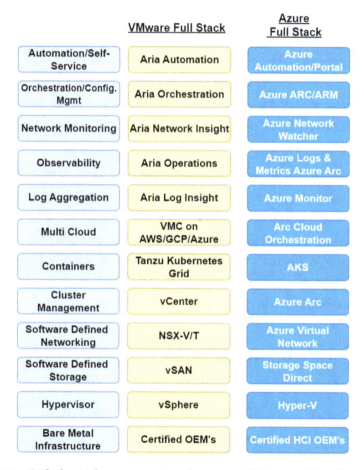

Figure 5-3. *Side by Side comparison between VMware Cloud Foundation and Microsoft Azure Stack HCI (or Azure Local) offering*

Please note that the VMware Cloud Foundation (VCF) is promoted as a purely on-premises offering. Whereas Microsoft Azure Stack HCI is promoted as a hybrid cloud offering where the management plan is Azure Portal, its associated products, and the backend infrastructure sites in your own datacenter.

Microsoft Azure Stack HCI

Microsoft Azure Stack HCI (hyper-converged infrastructure) is a hybrid cloud solution that extends Azure services and capabilities to on-premises environments. Designed to support modern workloads – such as virtual machines, containers, and microservices – Azure Stack HCI integrates seamlessly with Azure, enabling organizations to manage and deploy resources across their datacenters and the Azure cloud.

VMware Cloud Foundation

VMware Cloud Foundation is an integrated software platform that combines VMware's technology with software-defined networking and storage. This private cloud solution provides an automated, self-service infrastructure for deploying and managing workloads in a private datacenter. VMware Cloud Foundation can be deployed on-premises, offering organizations a customizable infrastructure environment tailored to their specific needs.

5.4.1 Key Comparisons

Table 5-2 provides the high-level comparison between VMware Cloud Foundation and Azure Stack HCI on different parameters.

Table 5-2. *Comparison between Azure Stack HCI and VMware Cloud Foundation*

Category	Vmware Cloud foundation (VCF)	Azure stack HCI
Architecture	Type-1 Hypervisor	Type-1 Hypervisor - Hyper-V at the core
	Light weight ESXi kernel that runs only from memory after boot	Windows 2022 have Azure stack HCI version of Operating system
Performance	Essentially Equal performance to Hyper-V or Azure stack HCI	Essentially Equal performance to Vmware
	ESXi 8 supports a maximum of 896 logical CPUs and upto 24 TB of RAM per host	Hyper-V supports up to 512 logical CPUs and up to 48 TB of RAM (windows server 2022) so as HCI
Usability	Built-in HTML5 web-based host management GUI	Azure portal as primary Management console. Also, Windows Admin Center (WAC), provides easy navigation.
	Vmware Cloud foundation (VCF) offers tools to build Private Cloud	Azure Portal integration offers Hybrid Cloud experience
Features	Aria suit for Automation, Orchestration, Operations, Log insights etc	Azure Portal, ARM teamplates and Azure tool set offers equivalent or better feature set.
	Tanzu Kubernetes Grid	Azure Kubernetes Service (AKS)
	NSX	Azure Virtual Networks
	VSAN	Storage Space Direct
	Guest VM's requires Additionl licensing	*With Azure hybrid benefits, the cost of HCI is $0 and Windows guest operationg systems runs "Free" if Hybrid benefit is leveraged.*
Scalability	ESXi capable of managing 1024 VMs and with vCenter it can support 96 hosts in a cluster, with Vmware vSAN it is 64 nodes in a cluster.	HCI supports at present maximum 1- 16 nodes cluster. With each node can run 1024 Virtual machines.
Cost	Recent anouncement with Broadcom removed access to free ESXi hypervisor	With Hybrid benefit, Azure stack HCI can run at no additional cost
	Vmware Cloud Foundation (VCF): MSRP $350/core == $5,600/16 core. Guest VM licenses are extra	With Azure Hybrid benefit and swap of Windows Datacenter server operating system allows unlimited Windows VM's at no extra license cost. Windows Datacenter MSRP cost is $6155 for 16 cores
		Without Hybrid benefit, cost of Azure stack HCI is $10/physical core/month Windows Server Subscription (for Guests) $23.3/Physical Core/month

5.4.1.1 Deployment Models

Azure Stack HCI: Primarily designed for hybrid cloud environments, Azure Stack HCI allows organizations to keep workloads on-premises while leveraging Azure services for backup, disaster recovery, and additional resources. This flexibility makes it suitable for businesses looking to retain some level of control over their data while benefiting from cloud capabilities.

VMware Cloud Foundation: As a private cloud solution, VMware Cloud Foundation focuses on enabling enterprises to build and run their own cloud environments within their data centers. It does not inherently offer hybrid capabilities but can be integrated with VMware's public cloud solutions for businesses that wish to extend their environment.

5.4.1.2 Use Cases

Azure Stack HCI: Organizations aiming to modernize their infrastructure while maintaining control over their data will find this solution particularly beneficial. Common applications include hybrid application deployments, managing edge workloads, and utilizing Azure services for real-time analytics and artificial intelligence.

A prominent application of Azure Stack HCI is in the area of virtual desktop infrastructure (VDI). With the rising need for remote work solutions, companies are adopting Azure Stack HCI to establish a scalable and effective VDI environment. This infrastructure allows organizations to deploy Windows desktops and applications virtually, ensuring that employees enjoy a seamless experience regardless of their location, all while upholding data security and compliance standards.

Additionally, Azure Stack HCI plays a crucial role in edge computing. As sectors such as manufacturing, retail, and healthcare increasingly adopt IoT and real-time data processing, Azure Stack HCI emerges as an optimal platform for data processing and analysis at the edge. Its capability to integrate with Azure services enables organizations to run applications closer to data generation points, thereby reducing latency and facilitating real-time insights. This functionality is vital for applications like predictive maintenance in manufacturing or real-time customer analytics in retail, where prompt decision-making can significantly improve operational efficiency and service delivery.

Another significant application of Azure Stack HCI is in the realm of disaster recovery and business continuity planning. This platform provides a resilient foundation that organizations can employ to establish effective backup and recovery strategies. By facilitating the replication of on-premises workloads to Azure, businesses can guarantee the availability of essential applications during outages or disasters, thus reducing both downtime and the risk of data loss. This integration not only improves the overall dependability of IT systems but also simplifies adherence to regulatory standards concerning data protection. With Azure Stack HCI, organizations can take advantage of scalable and cost-efficient high availability and disaster recovery solutions, making it an essential component of contemporary IT infrastructure.

VMware Cloud Foundation: Best suited for enterprises heavily invested in VMware technology that require a robust, fully integrated private cloud. Use cases often include mission-critical applications, large databases, and environments requiring strict compliance and security measures. One of the most significant use cases for VMware

Cloud Foundation is in the rapid deployment of private clouds. VCF offers a fully automated deployment process that allows organizations to provision and manage their cloud infrastructure quickly and efficiently. With its integrated components – such as VMware vSphere, vSAN, and NSX – VCF enables IT teams to easily create cloud environments tailored to their specific business needs.

Another prominent use case for VMware Cloud Foundation is that it integrates with VMware Tanzu, enabling developers to build, run, and manage applications across cloud environments with Kubernetes orchestration. This integration allows businesses to leverage the flexibility and scalability of containers while maintaining the robustness of their existing VMware infrastructure.

5.4.1.3 Performance and Scalability

Azure Stack HCI: Hyper-V serves as the foundational technology, complemented by Microsoft's Azure services to enhance scalability. This enables organizations to dynamically adjust their workloads in response to varying demand. The system is distinguished by its optimized performance for virtual machines and storage while also facilitating seamless integration with Azure for cloud bursting. Additionally, it takes advantage of other features, including disaster recovery and the comprehensive offerings available through Azure.

VMware Cloud Foundation: Having VMware ESXi hypervisor at its core, VCF offers better performance for enterprise applications by utilizing VMware's vSAN technology and providing advanced capabilities like workload balancing, automated scaling, and high availability. The proprietary technology serves well in scenarios with predictable workloads. The capacity and performance management are the ownership of the organizations implementing private cloud using VCF.

5.4.1.4 Management and Governance

Azure Stack HCI: Integrated with Azure management tools, Azure Stack HCI benefits from Azure Arc for unified governance across multi-cloud environments. This means organizations can enforce policies, monitor performance, and manage resources from a single pane of glass.

VMware Cloud Foundation: Provides a comprehensive management framework with Aria (previously called vRealize) Suite for automation, monitoring, and analytics. Organizations can leverage advanced capabilities to automate workflows and manage resources effectively, albeit primarily within their private cloud environment.

5.4.1.5 Cost Considerations

Azure Stack HCI: The cost is generally associated with the underlying hardware and Azure subscriptions for services consumed. This model can be cost-effective for organizations that already leverage Azure, as it allows for a pay-as-you-go approach for cloud services.

VMware Cloud Foundation: The initial expenses are generally elevated because of the investment required for VMware licenses and compatible hardware. Broadcom provides a subscription model; however, recent business strategies from Broadcom indicate a desire to impose longer-term commitments and contracts, typically spanning three years or more. Organizations may experience a decrease in total cost of ownership (TCO) over time by optimizing resource utilization and consolidating technologies to align with Broadcom's native offerings, provided that their strategic plans allow for continued engagement with a single vendor.

When evaluating Microsoft Azure Stack HCI versus VMware Cloud Foundation, organizations should take into account their unique needs, current technology investments, and overarching cloud strategy. Azure Stack HCI is particularly strong in hybrid environments, offering an efficient method for cloud integration, whereas VMware Cloud Foundation functions as a flexible private cloud solution. The final choice will depend on various elements, including the necessity for real-time scalability, the types of workloads, management preferences, and financial constraints.

By thoroughly assessing these two prominent cloud technologies, companies can better prepare themselves for success in an increasingly dynamic digital environment.

5.5 Summary

This chapter provides a comprehensive examination of Microsoft Hyper-V, tracing its historical evolution from its inception as a virtualization solution to its current standing in the industry. Initially introduced in 2008 as a component of Windows Server, Hyper-V has undergone significant improvements and expansions, solidifying its presence as a key player in enterprise virtualization. The narrative details its milestones and enhancements, highlighting how it has adapted to meet the demands of modern data centers and cloud environments.

A comparative analysis between Hyper-V and VMware further enriches the chapter, showcasing their respective strengths and weaknesses. While VMware has long been a dominant force in virtualization, Hyper-V offers appealing integration with Microsoft ecosystems and cost-effective licensing options. This section underscores the advantages and disadvantages of both platforms, giving readers critical insights into their functionalities and operational efficiencies.

The architecture of Hyper-V is examined in depth, shedding light on its core components and design principles that enable robust virtualization. The discussion reveals how Hyper-V utilizes a microkernel design to improve performance, scalability, and security, distinguishing it from traditional hypervisor architectures.

Furthermore, the chapter deconstructs hyper-converged infrastructure (HCI) architecture, with a specific emphasis on Azure Stack HCI. Readers are introduced to Azure Stack HCI's innovative framework, which merges compute, storage, and networking into a single integrated solution, optimizing resource management and deployment efficiency. The chapter highlights key features of Azure Stack HCI, such as its seamless integration with Azure services, increased agility, and centralized management capabilities.

Finally, the feature comparison between Azure Stack HCI and VMware Cloud Foundation presents a critical perspective on their functionalities. While both solutions offer robust capabilities for modern workloads, the chapter delineates specific areas where Azure Stack HCI may excel, such as simplicity in hybrid cloud deployments and advanced monitoring tools. This comparative analysis aims to equip readers with the information needed to make informed decisions regarding their infrastructure strategy in the evolving landscape of cloud computing and virtualization.

CHAPTER 6

Nutanix: The Challenger's Approach

As the digital environment continues to transform rapidly, with virtualization and cloud infrastructure at its foundation, organizations are confronted with critical choices that could influence their long-term IT strategies. VMware has maintained its leading position in this domain for many years; however, the increasing demand for flexibility, scalability, and cost-effectiveness is prompting companies to explore alternatives beyond traditional vendors. This transition has paved the way for innovative solutions like Nutanix, which has emerged as a significant contender through its groundbreaking approach to hyper-converged infrastructure (HCI).

In this chapter, we will thoroughly examine the reasons Nutanix has garnered the interest of CXOs, technologists, and decision-makers seeking transformative options. Nutanix's architecture departs from conventional models, delivering a software-defined infrastructure that seamlessly combines compute, storage, and networking into a unified platform. This approach offers a level of agility and simplicity that traditional virtualization solutions often find challenging to achieve. We will analyze Nutanix's distinct value proposition, focusing on essential aspects such as cost management, operational efficiency, and scalability, while providing a comparative analysis with VMware, the long-standing leader in the virtualization arena.

© Sumit Bhatia and Chetan Gabhane 2025
S. Bhatia and C. Gabhane, *Navigating VMware Turmoil in the Broadcom Era*,
https://doi.org/10.1007/979-8-8688-1264-4_6

For decision-makers, the conversation surrounding the transition from VMware to Nutanix transcends merely selecting a new platform; it involves developing a future-ready IT strategy that aligns with changing business goals. While VMware continues to deliver robust enterprise-grade solutions, Nutanix offers a novel perspective by minimizing complexity, lowering costs, and enabling a multi-cloud and hybrid cloud strategy with seamless integration across AWS, Azure, and Google Cloud.

We will evaluate the advantages of implementing Nutanix, focusing on its ability to simplify management, lower operational costs, and improve cost efficiency, especially in relation to cloud-native applications and workloads. The discussion will center on Nutanix's innovative strategies, particularly its effectiveness in mitigating security threats through comprehensive, built-in security measures. Furthermore, we will underscore the streamlined 1-click UI upgrade process, which boosts operational efficiency and minimizes downtime.

6.1 The History of Nutanix: Redefining the Virtualization Landscape

Nutanix has emerged as a key player in the virtualization and cloud infrastructure landscape, characterized by its innovative spirit, disruptive strategies, and an unwavering commitment to revolutionizing modern data center operations. Established in 2009, Nutanix has taken a daring approach, challenging established industry giants and expanding the horizons of hyper-converged infrastructure (HCI). For organizations looking for a viable alternative to conventional virtualization solutions such as VMware, Nutanix offers an attractive option that combines ease of use with robust capabilities, delivering a comprehensive software-defined infrastructure platform that seamlessly integrates compute, storage, networking, and virtualization.

At the heart of Nutanix's value proposition is its capacity to convert intricate, fragmented infrastructure into a cohesive, scalable solution that meets both cloud readiness and enterprise standards. This strategy goes beyond merely replacing existing virtualization technologies; it aims to fundamentally reshape how organizations perceive and manage their IT infrastructure. Nutanix's focus on enhancing operational efficiency, scalability, and minimizing complexity has positioned it as a prominent choice in the hyper-convergence market.

6.1.1 Early Beginnings: A Vision to Simplify Infrastructure

Nutanix emerged from the necessity to streamline the complexities inherent in traditional IT infrastructure. In its formative years, enterprise IT landscapes were typically marked by distinct systems for storage, computing, and networking, each necessitating specialized teams for operations and support. The founders of Nutanix identified this inefficiency and envisioned a solution that would integrate these functions into a unified, software-defined platform. They launched Nutanix as a hyper-converged infrastructure (HCI) solution designed to eliminate the intricacies of hardware management, enabling organizations to concentrate on deploying applications and services rather than overseeing the foundational infrastructure.

From the beginning, Nutanix adopted a distinctive approach. Rather than merely offering software to virtualize hardware, Nutanix developed a platform that amalgamated computing, storage, and networking into a cohesive unit, all governed by its proprietary Nutanix Acropolis operating system (AOS). This innovation resulted in an infrastructure stack that was not only simpler to manage but also inherently scalable, empowering businesses to expand their capacity as required without the complications typically associated with scaling traditional infrastructure.

6.1.2 The Birth of Acropolis: 2015

The introduction of Nutanix Acropolis in 2015 represented a pivotal moment in the company's evolution. Acropolis served as Nutanix's response to conventional hypervisors such as VMware ESXi and Microsoft Hyper-V, providing a cohesive, cloud-native virtualization solution that was designed to integrate effortlessly with the Nutanix HCI platform. By developing its own hypervisor, Nutanix sought to empower organizations with greater control over their environments while minimizing reliance on external virtualization vendors. Acropolis not only featured a native hypervisor but also supported integration with other hypervisors, allowing businesses the flexibility to select the most suitable solution for their requirements.

Additionally, Acropolis introduced Nutanix Prism, a user-friendly management interface that streamlined the administration of the entire infrastructure stack. Prism's single-pane-of-glass design enabled IT teams to oversee compute, storage, networking, and virtualization through a consolidated interface, enhancing operational efficiency and reducing administrative burdens. This emphasis on simplicity and operational effectiveness has been a significant differentiator for Nutanix and continues to be a core aspect of its value proposition.

6.1.3 Expanding Capabilities: AHV and Hybrid Cloud Integration

Nutanix's evolution continued beyond Acropolis. As cloud technology gained prominence in enterprise IT, Nutanix adapted to fulfill the evolving demands of its clientele. The launch of Nutanix AHV (Acropolis Hypervisor) solidified its status as a competitive alternative to conventional hypervisors. AHV was meticulously crafted to be a lightweight, integrated hypervisor that operates flawlessly on the Nutanix

HCI platform. In contrast to other hypervisors, which typically serve as supplementary components to existing systems, AHV was specifically engineered for hyper-converged infrastructures, delivering enhanced performance and streamlined management.

Moreover, Nutanix expanded its cloud functionalities, introducing features that facilitate seamless integration of on-premises systems with public cloud services such as AWS, Microsoft Azure, and Google Cloud. By prioritizing hybrid cloud integration, Nutanix empowered organizations to extend their infrastructure beyond traditional data centers, granting them the versatility to deploy applications and services in the most suitable environments – whether on-premises, in the cloud, or in a hybrid setup.

6.1.4 The Present and Future: Nutanix in a Cloud-First World

Nutanix remains at the forefront of innovation, spearheading advancements in cloud-native infrastructure and redefining possibilities in a cloud-first environment. As organizations increasingly adopt multi-cloud and hybrid cloud strategies, Nutanix has established itself as a vital facilitator of this evolution. Its capacity to deliver a cohesive, unified infrastructure platform that integrates both on-premises and cloud settings makes it a preferred option for companies aiming to secure their IT investments for the future.

Looking forward, Nutanix is dedicated to enhancing its cloud-native offerings, particularly in areas such as containerization, microservices, and edge computing. The growing prominence of Kubernetes and the transition to containerized workloads have prompted Nutanix to create solutions that cater to these contemporary application frameworks, ensuring that businesses can operate their applications effectively, irrespective of their deployment locations.

In a cloud-first landscape, the ability to seamlessly connect with public cloud services while retaining oversight of on-premises systems will be essential for organizations striving to maintain a competitive edge. Nutanix's focus on adaptability, scalability, and ease of use positions it as a significant contender in this dynamic environment, equipping businesses with the necessary tools to manage their infrastructure efficiently and prepare for the future.

In summary, Nutanix has established a distinctive position within the virtualization and cloud infrastructure sector. Its commitment to hyperconvergence, along with its integrated AHV hypervisor and robust hybrid cloud functionalities, presents a formidable alternative to conventional solutions like VMware. For organizations looking to simplify operations, reduce expenses, and embrace a multi-cloud future, Nutanix provides an attractive solution that merges the advantages of on-premises control with cloud scalability.

6.2 Nutanix Architecture and Value Proposition

As organizations face the increasing challenges of managing contemporary IT infrastructure, the need for solutions that streamline operations while boosting performance has reached unprecedented levels. Traditional virtualization platforms like VMware have historically served as the foundation for enterprise IT environments, providing dependable virtualization capabilities. However, with the swift advancement of cloud computing, multi-cloud strategies, and hybrid setups, many companies are exploring alternatives that deliver greater flexibility, cost-effectiveness, and scalability without sacrificing performance.

This is where Nutanix comes into play. Nutanix has transformed the virtualization landscape with its cutting-edge hyper-converged infrastructure (HCI), which combines compute, storage, and networking into a cohesive software-driven platform. In this section, we will delve into the architecture that supports Nutanix's HCI solution, highlighting how it contrasts with traditional configurations and why it has emerged as an attractive choice for businesses looking to transition from VMware. Additionally, we will examine the key advantages Nutanix offers, particularly regarding cost, performance, ease of management, and cloud integration. Please refer to Figure 6-1, which shows a high-level overview of how hyper-converged infrastructure is organized into distinct layers and interconnected at the software level.

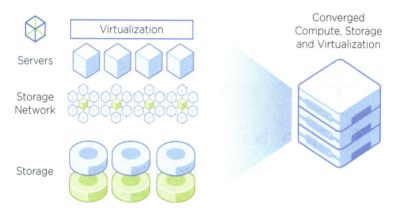

Figure 6-1. *Nutanix hyper-converged solution overview*

Nutanix's architecture, shown in Figure 6-2, exemplifies the seamless integration of various components, functioning like a meticulously crafted machine where each element is essential, working in concert to enhance the overall power and efficiency of the system. Envision it as a meticulously orchestrated ensemble, where each instrument contributes significantly to a unified and melodious performance. The real power of Nutanix lies not only in the exceptional attributes of its individual elements but also in their synergistic interactions, which enhance one another's capabilities to create a seamless, high-efficiency infrastructure.

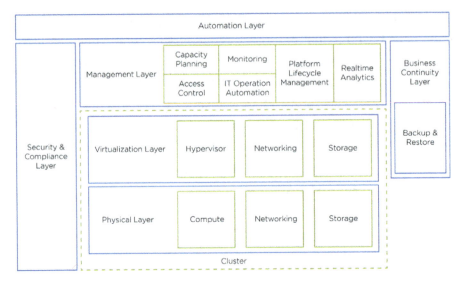

Figure 6-2. *Nutanix's architectural overview*

6.2.1 The Nutanix Cluster: A Unified Foundation

The journey commences with the Nutanix cluster. Each node within this cluster functions as a collaborative team member, combining their computing capabilities and storage resources. Envision them as a collective of experts; while each node possesses individual strengths, their collaboration results in a system that far exceeds the capabilities of its individual components. Through effective communication, these nodes establish a cohesive cluster that can efficiently distribute workloads, manage resources, and maintain uninterrupted availability, even in the event of a node failure. This interaction among nodes is fundamental to the renowned resilience of Nutanix. They continuously exchange information, ensuring data replication, workload distribution, and resource accessibility as required. Please refer to Figure 6-3 for the VMware versus Nutanix stack, which provides a high-level overview of different stacks and how they're optimized in Nutanix.

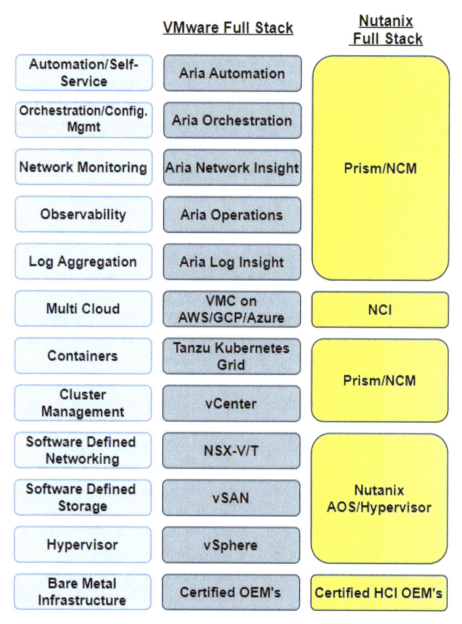

Figure 6-3. VMware versus Nutanix stack

When evaluating VMware and Nutanix, it is crucial to recognize their unique strategies regarding cloud infrastructure and management, as demonstrated by their respective offerings: Nutanix Cloud Infrastructure (NCI) and VMware Cloud Foundation (VCF). Nutanix presents itself as a cohesive hybrid multi-cloud platform that effectively integrates data centers, cloud settings, and edge deployments. In contrast, VMware utilizes distinct platforms designed for on-premises and VMware-managed cloud environments on AWS and Azure, which can lead to increased complexity for users.

Nutanix features an intuitive, HTML5-based management interface that prioritizes ease of use, offering one-click upgrades for software, hypervisor, and firmware, thereby significantly simplifying operations. Conversely, VMware necessitates multiple management interfaces based on the deployment environment, which can complicate user experience. Additionally, Nutanix provides its AHV hypervisor at no extra charge while also supporting VMware ESXi for certain scenarios. VMware predominantly depends on ESXi as its standard hypervisor, which contributes to higher licensing expenses for users.

In terms of scalability, Nutanix stands out with its node-based architecture that automatically reallocates data as clusters grow. While VMware's scalability is strong, it is often limited by specific vSAN configurations and minimum node requirements, which can restrict flexibility. Regarding cloud management, Nutanix Cloud Management (NCM) offers a centralized suite for managing multi-cloud environments, application migrations, and automation. In comparison, VMware's vRealize Suite, although comprehensive, requires separate licenses and setup processes, which can increase operational complexity.

Data localization in Nutanix guarantees that virtual machine storage is maintained on the same host, which enhances performance and minimizes dependence on network bandwidth. In contrast, VMware's methodology, while effective, does not match this level of efficiency, particularly in environments where bandwidth is limited. From a security perspective, Nutanix incorporates micro-segmentation and comprehensive disaster

256

recovery capabilities, making it easier to address potential security breaches. Although VMware also provides robust security features, their implementation across multi-cloud settings can be more intricate.

Cost factors further emphasize Nutanix's benefits, as its integrated hypervisors and efficient upgrade processes lead to lower operational costs. Conversely, VMware's dependence on additional licenses for its tools and hypervisors often results in higher initial and ongoing expenses.

Nutanix excels in its simplicity, operational efficiency, and adaptability, positioning it as an ideal solution for hybrid and multi-cloud environments. Conversely, VMware's strong integration with traditional enterprise systems and its support for containerized applications are well-suited for enterprises with larger budgets and uniform infrastructure requirements. This comprehensive comparison highlights the unique advantages of each platform, allowing organizations to make well-informed choices based on their individual needs and priorities.

6.2.2 Acropolis Hypervisor (AHV): The Heartbeat of Virtualization

At the heart of Nutanix's operations lies the Acropolis Hypervisor (AHV). Rather than merely existing atop the Nutanix cluster, AHV is intricately woven into its very structure. This hypervisor engages directly with the nodes, orchestrating workloads and guaranteeing that virtual machines (VMs) are efficiently deployed and managed throughout the cluster. AHV transcends the role of a typical hypervisor; it functions as the central nervous system of the Nutanix ecosystem, seamlessly interfacing with both compute and storage layers to ensure intelligent resource allocation. Upon the deployment of a new VM, AHV automatically identifies the most suitable node for hosting, consults the storage layer for optimal data placement, and balances workloads to prevent congestion. It operates as if it possesses a comprehensive awareness of the system, continuously optimizing performance.

6.2.3 Distributed Storage Fabric (DSF): The Lifeline of Data

The true innovation of Nutanix emerges with its Distributed Storage Fabric (DSF), a cornerstone of its architecture, seamlessly integrated with AHV. While AHV efficiently manages virtualization tasks, DSF ensures that all storage across the nodes is consolidated into a single, cohesive pool. As AHV initiates virtual machine deployments, DSF determines the optimal storage location for data, prioritizing ultra-low latency by placing data as close to the application as possible. This strategic placement minimizes delays and enhances application performance.

Simultaneously, DSF guarantees redundancy by replicating data across nodes, ensuring high availability. The elegance of DSF lies in its ability to manage data movement seamlessly, akin to a well-orchestrated traffic system that prevents congestion and bottlenecks. Its operations are fluid, with DSF consistently providing AHV with updates on the storage layer's health and performance to ensure optimal operation.

Consider DSF as the circulatory system of Nutanix: data flows effortlessly through the infrastructure, perpetually available, secure, and replicated. This synergy ensures that computational processes remain unaffected by storage constraints, making Nutanix a robust solution for enterprises demanding performance, reliability, and simplicity.

6.2.4 Prism: The All-Seeing Eye

Prism, the management console from Nutanix, serves a purpose far beyond that of a mere control panel. It interfaces with every element of the system – be it the cluster, AHV, or DSF – offering a streamlined and intuitive user experience. Acting as the central nervous system of the infrastructure, Prism continuously gathers information from all components, processes it, and presents administrators with a clear, real-time overview of operations.

The Nutanix Prism management console acts as a robust orchestrator, issuing commands throughout the infrastructure while consistently gathering updates from all components to ensure optimal performance. Designed for an exceptional user experience, Prism emphasizes simplicity and intuitiveness, necessitating minimal training for administrators. This approach allows for effortless monitoring and management of the entire IT infrastructure. Applying updates and upgrades is as straightforward as updating a smartphone application, reflecting Nutanix's dedication to operational simplicity. In comparison to other platforms, such as VMware's management tools, Nutanix Prism is frequently noted for requiring significantly less training – up to five times less – enabling teams to quickly utilize its features without extensive learning curves. By focusing on user-friendliness, Prism boosts both productivity and operational efficiency in contemporary IT environments.

6.2.5 NCM: The Automator of Interactions

Nutanix has renamed its automation engine, previously known as Calm, to Nutanix Cloud Manager (NCM) Self-Service. This platform enhances automation by managing interactions among various elements within hybrid cloud environments. The deployment of a multi-tier application usually involves several steps across different systems. NCM Self-Service simplifies this process by orchestrating workflows throughout the infrastructure, which includes virtual machine provisioning, storage management, and deployment oversight.

By converting intricate operations into simple, repeatable templates, NCM Self-Service facilitates effective application management. After deployment, it continues to oversee and enhance performance. For example, if a virtual machine experiences high demand or low usage, NCM Self-Service works with the underlying infrastructure to redistribute resources, ensuring maximum efficiency. Acting like a conductor in an orchestra, it not only leads the performance but also adjusts the instruments in real time to maintain seamless harmony.

This rebranding underscores Nutanix's dedication to delivering a cohesive platform that streamlines application management across hybrid clouds through self-service capabilities, automation, and centralized role-based governance.

6.2.6 ADSF: The Bridge Between Compute and Storage

AHV manages virtualization while DSF oversees storage; however, the true integration occurs within the Acropolis Distributed Storage Fabric (ADSF). ADSF plays a crucial role in unifying compute and storage resources, eliminating the traditional separation between them. Picture compute and storage as two synchronized dancers, their movements harmonized to deliver rapid response times and enhanced workload efficiency.

When a new virtual machine is initiated, AHV collaborates with ADSF to address both compute and storage requirements concurrently. This process eliminates delays and the need for sequential adjustments in storage resources. ADSF effectively connects compute and storage, ensuring they operate in harmony, thereby optimizing performance without necessitating continuous management.

6.2.7 Data Protection and Security: The Silent Guardian

Nutanix excels in managing compute and storage with effortless communication while also providing robust protection behind the scenes. Its integrated data protection features interact directly with the Distributed Storage Fabric (DSF) to guarantee that data is consistently replicated and accessible. In the event of a node failure, DSF ensures that data remains available on other nodes, all without causing any disruption. This level of interaction and redundancy equips Nutanix to handle unforeseen circumstances, delivering a security layer that functions quietly in the background.

The synergy among components such as AHV, DSF, Prism, Calm, and ADSF is what truly enhances Nutanix's capabilities. Each element operates not in isolation but in constant communication, ensuring that compute, storage, automation, and management function as a unified system. The outcome is an infrastructure that is not only straightforward to manage but also remarkably adaptable and resilient, capable of responding to changes and demands seamlessly. Nutanix's architecture emphasizes collective functionality, creating a system well-equipped to meet the challenges of today's IT environment.

6.2.8 Nutanix Architecture: Building a Simplified and Scalable Foundation

Nutanix's innovation is fundamentally driven by its hyper-converged infrastructure, which removes the necessity for distinct, isolated hardware for storage, computing, and networking. Nutanix DSF comes bundled with the Nutanix platform, eliminating separate licensing costs for storage. This results in more predictable pricing compared to VMware's vSAN per-core mode. The software-defined platform from Nutanix integrates these traditionally separate elements into a cohesive system, allowing organizations to oversee their entire infrastructure through a single, unified interface.

VMware's vSAN is closely integrated with VMware products such as vSphere and NSX, making it an ideal option for organizations that are already part of the VMware ecosystem. However, its pricing model, which is based on a per-core basis, can become excessively costly in environments with a high number of cores, leading to significant expenses when scaling. Furthermore, integrating vSAN into non-VMware environments can be challenging, which may restrict its adaptability in diverse infrastructure settings.

Conversely, Nutanix's architecture is founded on its powerful Acropolis Operating System (AOS), which drives the platform and offers advanced features like data deduplication, compression, and erasure coding to improve storage efficiency. Central to Nutanix's design is the Distributed Storage Fabric (DSF), which aggregates storage resources from all nodes within a cluster, facilitating effortless scaling and efficient data management. This forward-thinking approach enables organizations to transcend the constraints of traditional SAN and NAS solutions, which tend to be more expensive, complex, and less flexible in meeting dynamic scaling requirements. Nutanix's offering is particularly advantageous for enterprises that prioritize agility, simplicity, and cost-effectiveness in their storage solutions.

Moreover, Nutanix's hyper-converged infrastructure is designed with resilience and fault tolerance as core principles. Featuring self-healing capabilities, the system can automatically redirect workloads in the event of hardware failures, ensuring minimal downtime and continuity of operations. For enterprises relying on mission-critical applications, this high degree of reliability offers significant value, as it mitigates the risk of disruptions and boosts overall performance.

6.2.9 Nutanix's Value Proposition: Cost, Performance, and Simplicity

For Nutanix, it presents a compelling value proposition for decision-makers, particularly in terms of cost efficiency, performance enhancement, and ease of management. Its architecture enables organizations to streamline their IT infrastructure, minimizing reliance on various hardware vendors and disparate systems. This consolidation results in lower capital expenditure (CapEx) and operational expenditure (OpEx), as there is no longer a need to maintain separate systems for storage, computing, and networking.

From a performance standpoint, Nutanix's web-scale architecture facilitates the distribution of workloads across multiple nodes within a cluster, effectively eliminating bottlenecks and improving overall system performance. The distributed storage architecture is crucial in this regard, offering high throughput and low-latency data access, which is vital for performance-critical applications like databases and virtual desktop infrastructure (VDI).

Additionally, Nutanix's software-centric approach significantly reduces the manual configuration and management typically associated with traditional infrastructure. IT teams can utilize automation and orchestration tools available within the Nutanix platform to simplify routine tasks such as provisioning, scaling, and patch management. This not only alleviates the administrative workload for IT personnel but also speeds up the deployment of new applications and services.

For organizations aiming to implement a multi-cloud or hybrid cloud strategy, Nutanix provides seamless cloud integration through Nutanix Clusters, which extend the platform into public cloud environments like AWS and Microsoft Azure. This capability allows businesses to operate workloads across both on-premises and cloud settings with consistent management and orchestration, offering the necessary flexibility to respond to evolving business needs.

6.2.9.1 Nutanix Versus VMware: Key Differentiators

When evaluating Nutanix against VMware, several significant distinctions emerge, positioning Nutanix as a compelling choice for organizations aiming to upgrade their infrastructure. A primary advantage is the cost-effectiveness of Nutanix's platform. VMware's licensing structure, especially for its vSphere and vSAN offerings, can become excessively costly, particularly for enterprises with extensive deployments. In contrast, Nutanix's comprehensive pricing model removes many of the licensing expenses tied to external hypervisors and storage solutions, resulting in a more predictable and economical pricing framework.

Regarding management ease, Nutanix's Prism console offers a more user-friendly and efficient experience compared to VMware's vCenter, which often necessitates specialized knowledge for effective management. The capability to oversee compute, storage, and networking through a unified interface lessens the learning curve for IT personnel and facilitates quicker problem resolution.

Nutanix also stands out in terms of flexibility and scalability. While VMware provides robust virtualization features, Nutanix's web-scale architecture is inherently more adaptable, enabling organizations to incorporate additional nodes into their infrastructure with minimal disruption. This scalability is especially advantageous for businesses with fluctuating workloads or those expecting rapid expansion, as it allows for resource scaling as required without necessitating significant infrastructure changes.

6.2.9.2 Security and Data Protection

In today's business environment, security has emerged as a critical priority for decision-makers, and Nutanix effectively meets this challenge with a range of integrated security features. The platform offers native data encryption for both data at rest and in transit, ensuring the protection of sensitive information across the entire infrastructure. Additionally, Nutanix's microsegmentation capabilities empower IT teams to establish precise security policies that isolate and safeguard specific workloads, thereby reducing the risk of lateral attacks within the data center. The self-healing architecture of Nutanix enhances the platform's security and resilience; in the event of hardware failures or security breaches, the distributed system automatically reroutes workloads and guarantees data replication across multiple nodes, thus minimizing downtime and mitigating the effects of such incidents. Nutanix also facilitates cross-hypervisor disaster recovery (DR), allowing for smooth failover and failback between ESXi and AHV clusters, which improves flexibility and the overall efficiency of disaster recovery efforts. Furthermore, Nutanix provides comprehensive disaster recovery solutions, including Nutanix NCM for automation and Nutanix Xi Leap for Disaster Recovery as a Service (DRaaS), equipping organizations with vital tools to ensure business continuity during unexpected events. For a concise overview, please refer to Figure 6-4 that highlights these key aspects.

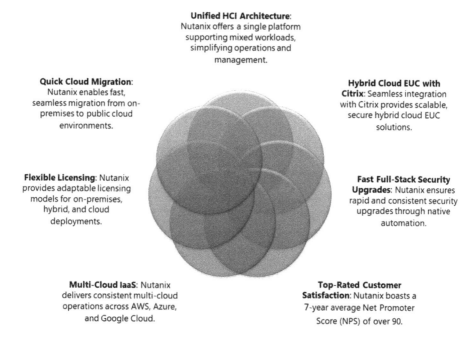

Figure 6-4. *Why Nutanix as VMware alternative*

6.3 Cost, Management, and Suitability Comparison with VMware

When evaluating Nutanix against VMware, the choice ultimately hinges on the unique requirements, growth objectives, and IT aspirations of the organization. Nutanix distinguishes itself through its provision of cost-efficient, scalable, and highly automated solutions that simplify operations while offering the necessary flexibility in today's hybrid and multi-cloud environments. Its integrated management capabilities, versatility across diverse applications, and reduced operational expenses position it as a strong contender to VMware, especially for organizations that value agility, ease of use, and readiness for the future. For decision-makers, Nutanix presents a strategic option that addresses contemporary IT challenges, empowering businesses to innovate while maximizing their infrastructure investments. Let us explore specific criteria that assist organizations and decision-makers in selecting the most suitable solution with future considerations in mind.

6.3.1 Cost Efficiency and TCO (Total Cost of Ownership)

At first glance, the initial investment in Nutanix may seem similar to that of VMware; however, the true financial benefits become evident over time through total cost of ownership (TCO) and lower operational expenses. Nutanix's hyper-converged infrastructure (HCI) integrates compute, storage, and networking into a unified platform, thereby eliminating the necessity for separate, often expensive, storage area networks (SANs) and minimizing hardware requirements. This consolidation leads to fewer components, reduced maintenance needs, and decreased energy usage. For organizations, this streamlined physical infrastructure results in substantial long-term savings in data center operations, including costs associated with power, cooling, and space.

Nutanix considers infrastructure and virtualization to be fundamental components that should be straightforward and manageable. To actualize this vision, Nutanix has seamlessly incorporated its proprietary hypervisor, AHV, into the platform. AHV delivers comparable functionality to VMware's Enterprise Plus but without any additional charges, allowing organizations to access advanced virtualization capabilities without incurring extra costs. This approach removes the typical expenses linked to virtualization software, thereby enhancing long-term savings and operational efficiency.

In contrast, VMware typically necessitates investments across multiple infrastructure layers – servers, storage, and networking – each of which may incur distinct licensing, maintenance, and hardware expenses. The licensing structures in VMware environments can also be inflexible, often resulting in lock-in situations that bind businesses to costly, incremental upgrades. Nutanix, on the other hand, offers a more adaptable licensing approach and a pay-as-you-grow model, allowing organizations greater control over their budgets. This flexibility enables them to expand their environments without the constant worry of licensing limitations or unforeseen expenses.

6.3.2 Management Complexity and Operational Simplicity

From a management standpoint, Nutanix greatly simplifies operations by providing a cohesive management interface. The native management console, Nutanix Prism, offers a user-friendly, consolidated view of the entire infrastructure. IT teams can oversee virtual machines (VMs), storage, and network resources all from a single interface, which streamlines daily tasks. In contrast, VMware's toolset can be quite fragmented, requiring the use of separate components such as vSphere, vSAN, and NSX, which often necessitate multiple consoles and specialized expertise, resulting in higher administrative burdens and a more challenging learning process.

Additionally, Nutanix's automation features enhance infrastructure management by minimizing the need for manual tasks, allowing IT teams to concentrate on more strategic projects. The platform's AI-driven tools automate processes like patch management, infrastructure upgrades, and troubleshooting, promoting consistency and reliability throughout the environment. While VMware is certainly powerful, it typically demands extra orchestration tools to reach a comparable level of automation, which can increase both complexity and expenses.

Furthermore, Nutanix's self-healing architecture boosts resilience by automatically identifying and rectifying system failures, ensuring continuous operations. For organizations in fast-paced environments, this degree of automation and operational ease can serve as a crucial advantage. Although VMware's systems are strong, they often require more manual effort and dependence on various third-party solutions to provide similar capabilities.

6.3.3 Suitability Across Different Use Cases

Nutanix's primary advantage lies in its versatility across diverse use cases. Its hyper-converged infrastructure is particularly suited for enterprises aiming for cloud-like responsiveness within their on-premises data centers, as well as for organizations pursuing a hybrid cloud model. Nutanix facilitates seamless integration with major public cloud providers such as AWS, Azure, and Google Cloud, allowing businesses to expand their workloads into the cloud without the risk of vendor lock-in. This adaptability enables organizations to migrate workloads between on-premises and cloud environments in response to evolving business demands.

In contrast, while VMware is recognized as a leader in virtualization, its cloud strategy, especially with VMware Cloud, tends to confine organizations within its ecosystem, resulting in less flexibility for cloud migration and a greater reliance on VMware-specific infrastructure.

Nutanix's agnostic cloud approach and multi-cloud capabilities empower organizations to implement a best-of-breed strategy, circumventing the constraints associated with a single-vendor cloud solution.

For edge computing scenarios, Nutanix's streamlined deployment models and simplified management processes are particularly beneficial. This allows businesses to effectively oversee remote sites, branch offices, and edge locations with minimal overhead, a task that can be more complicated in VMware environments due to their heavier infrastructure demands and more complex management requirements.

Nutanix also stands out in disaster recovery and business continuity. With its integrated disaster recovery solutions, Nutanix provides automated failover, replication, and recovery across both on-premises and cloud environments. Its built-in security features and self-healing capabilities ensure that organizations can maintain resilience without depending on multiple third-party solutions. Although VMware offers similar functionalities, they often necessitate additional licensing and integration with third-party vendors, which can introduce complexity and increase costs.

6.3.4 Future-Ready Infrastructure

Nutanix's architecture is fundamentally designed to be future-ready, allowing it to adapt seamlessly to new technologies as they emerge. By supporting both containers and Kubernetes in addition to traditional virtual machines, it enables organizations to implement modern application frameworks without disrupting their current operations. The cloud-native design of Nutanix emphasizes flexibility, empowering businesses to innovate and expand without being constrained by infrastructure challenges.

In contrast, VMware, despite its ongoing innovations, frequently launches new solutions that necessitate substantial upgrades or additional investments in other products. This can pose challenges for organizations aiming to remain agile and embrace the latest technologies without facing excessive costs.

Table 6-1 is a comprehensive comparison table between Nutanix and VMware, focusing on cost, management, and suitability.

Table 6-1. *Detailed comparison of Nutanix versus VMware on various parameters*

Parameter	Nutanix	VMware
Total Cost of Ownership (TCO)	Lower TCO due to integrated HCI model, reducing hardware and operational costs. Nutanix eliminates the need for separate compute, storage, and networking layers.	Higher TCO due to separate licensing costs for compute, storage, and networking. Requires multiple platforms for full infrastructure, increasing operational expenses.
Licensing Flexibility	Flexible, consumption-based licensing model, with options for subscription and perpetual licensing. Ideal for hybrid and multi-cloud deployments.	VMware has streamlined its licensing process; the transition to subscription models introduces increased recurring expenses, which may affect businesses aiming to control their budgets through initial investments.
Initial Setup Costs	Lower setup costs due to a simplified HCI solution, reducing infrastructure sprawl. Quick deployments help reduce time to value.	VMware's new vSAN licensing bundle, which is now based on a per-core pricing model, has made it more expensive compared to Nutanix's all-inclusive pricing.

(*continued*)

Table 6-1. (*continued*)

Parameter	Nutanix	VMware
Management Complexity	Simplified, single-pane-of-glass management with Nutanix Prism. Automation and one-click upgrades make day-to-day management easy.	More complex with different management tools for compute (vSphere), storage (vSAN), and networking (NSX). Requires more specialized expertise.
Cloud Integration	Seamless integration with major public clouds (AWS, Azure, Google Cloud) with consistent operations across environments. Nutanix Clusters offer easy extension to the cloud.	VMware Cloud integrates with AWS, Azure, Google Cloud, but can require additional tools like vCloud Director and separate management interfaces.
Scalability	High scalability with linear growth, allowing enterprises to add nodes without disruptive changes. Ideal for hybrid cloud and edge computing environments.	Scalable, but often requires larger upfront investments in specific areas like storage or networking, which can result in inefficiencies at scale.
Security Management	Nutanix offers integrated, full-stack security with automated updates, micro-segmentation, and compliance. Nutanix Flow provides built-in network security and governance.	VMware also provides strong security, but often through a patchwork of tools like NSX for micro-segmentation, and third-party integrations, making it more complex to manage.

(*continued*)

Table 6-1. (*continued*)

Parameter	Nutanix	VMware
Automation and Upgrades	One-click upgrades and extensive automation features reduce downtime and manual intervention. Supports self-healing and autonomous operations.	VMware requires more manual oversight during upgrades, especially across its different solutions (vSphere, NSX, and vSAN), often leading to higher administrative overhead.
Suitability for Hybrid Cloud	Designed natively for hybrid cloud with consistent operations across on-prem and cloud environments. Perfect for edge computing and distributed deployments.	VMware's hybrid cloud capabilities are robust but often require additional products and services, such as VMware Cloud Foundation, to match Nutanix's seamless integration.
Edge and Remote Sites	Nutanix excels in lightweight deployments, making it ideal for remote offices, branch offices (ROBO), and edge computing with minimal infrastructure with two nodes.	VMware vSAN 2-Node clusters support small-scale and remote office environments. It's ideal for ROBO deployments, but scalability may be limited based on workload demands.

This analysis illustrates the differences between Nutanix and VMware regarding cost, management, and overall suitability, assisting CXOs and decision-makers in selecting the most appropriate solution for their unique requirements. Nutanix generally provides a more straightforward and adaptable option with reduced ongoing expenses, whereas VMware's offerings are typically more comprehensive but may necessitate a greater financial commitment and increased management complexity.

6.4 Advantages and Disadvantages of Adopting Nutanix

Like any technology, the adoption of Nutanix comes with its own unique set of benefits and drawbacks that decision-makers and technologists responsible for guiding IT investment strategies must carefully evaluate. In this section, we will thoroughly explore these advantages and disadvantages, presenting them in a way that offers both clarity and insight into their long-term implications for an organization.

6.4.1 Advantages of Adopting Nutanix

As digital transformation progresses rapidly, organizations are increasingly in search of flexible, efficient, and scalable solutions for their infrastructure management. Nutanix, known for its hyper-converged infrastructure (HCI), has positioned itself as a viable alternative for those considering options beyond VMware.

6.4.1.1 Simplified Infrastructure Management

Consider a manufacturing enterprise with operations spanning several countries and numerous branch offices. In the past, each location was responsible for its own servers, storage solutions, and networking devices. This approach was not only intricate but also time-consuming and prone to errors. Nutanix simplifies these varied components into a cohesive system, enabling management of the entire infrastructure through a single interface – Nutanix Prism. This unified management perspective removes the necessity for various isolated management tools, thereby decreasing administrative burdens. IT administrators can efficiently deploy, monitor, and resolve issues within the infrastructure across a global network from a centralized platform. This is particularly advantageous for organizations

273

with streamlined IT teams, allowing them to concentrate on innovation instead of routine maintenance tasks. In essence, Nutanix transforms complexity into simplicity.

6.4.1.2 Cost Efficiency

For numerous CXOs, cost remains a pivotal consideration when assessing new technology solutions. Nutanix presents significant cost advantages compared to traditional three-tier architectures, which include servers, storage, and networking. By integrating these components into a software-defined framework, Nutanix removes the necessity for costly, specialized hardware, thereby lowering capital expenditures. Additionally, the use of commodity hardware allows organizations to sidestep vendor lock-in and secure more favorable pricing for future hardware acquisitions.

Take, for instance, a healthcare provider that transitioned from its outdated systems to Nutanix. This upgrade not only led to hardware cost reductions but also minimized their data center footprint, resulting in decreased power and cooling expenses. Furthermore, Nutanix facilitates predictable scaling, enabling organizations to expand gradually without incurring substantial upfront costs.

Seamless Cloud Integration

In today's landscape, where organizations are increasingly dependent on cloud solutions, Nutanix stands out by effectively integrating on-premises infrastructure with public cloud platforms. Regardless of whether a business utilizes AWS, Azure, or Google Cloud, Nutanix Clusters facilitate smooth transitions to and from the cloud, providing genuine hybrid cloud versatility.

Nutanix offers businesses the flexibility to extend their infrastructure to the public cloud through seamless integration with platforms like AWS, Azure, and Google Cloud. This allows organizations to scale resources dynamically, meeting increased demand during peak periods, and scale back after, only paying for what they use. By leveraging Nutanix's hybrid cloud capabilities, companies can maintain consistent performance and manage workloads across both private and public environments, offering cost efficiency and adaptability without the complexity of traditional infrastructure setups.

6.4.1.3 Enhanced Security

In the current landscape, cybersecurity is essential, and Nutanix addresses this by embedding security at every level, providing comprehensive protection from the ground up. Through Nutanix Flow, organizations can deploy micro-segmentation to isolate workloads and restrict threat movement. With a zero-trust security framework, Nutanix adds an extra layer of defense against unauthorized access. Additionally, the Nutanix hyper-convergence solution offers 800 automated security checks in line with Nutanix STIGs, covering both storage and virtualization. Any security violations are automatically detected and rectified, ensuring continuous protection.

This approach is especially beneficial for businesses like financial services, where sensitive customer data is securely separated from other workloads, mitigating the risk of breaches.

6.4.1.4 Scalability and Flexibility

Nutanix's architecture is built for scalability, making it an excellent choice for businesses expecting growth. Organizations can begin with a small deployment on a limited number of nodes and expand as necessary without interrupting their operations. In contrast to traditional infrastructure, which often demands substantial investment in additional hardware for scaling, Nutanix adapts to the business's needs, whether that involves increasing the number of virtual machines or enhancing storage capacity. For instance, a global logistics firm that ventured into new markets was able to rapidly scale its infrastructure using Nutanix, allowing each new office location to become operational with minimal setup time.

6.4.1.5 High Availability and Resilience

For companies, periods of downtime result in diminished revenue and eroded trust. Nutanix offers essential features like self-healing nodes and automated disaster recovery to enhance availability and resilience. In the event of a node failure, Nutanix seamlessly reallocates workloads to other nodes, ensuring continuous system functionality. Additionally, Nutanix's disaster recovery service, Nutanix Leap, enables businesses to test their disaster recovery plans in a network bubble without interrupting production. This ensures that recovery time objectives (RTO) and recovery point objectives (RPO) are consistently met, which is critical for sectors like healthcare and finance that demand uninterrupted service.

6.4.2 Disadvantages of Adopting Nutanix

While Nutanix offers numerous benefits, it's also important to recognize its potential drawbacks. Understanding these limitations will help organizations make a balanced decision.

6.4.2.1 Initial Learning Curve

Nutanix's architecture, while straightforward, significantly diverges from conventional IT systems. Organizations heavily reliant on legacy systems may face an initial adjustment period. IT personnel may need targeted training to grasp the platform's features and optimal usage strategies. Consider a government agency that has relied on outdated systems for years. Moving to Nutanix would demand not only a technological transition but also a cultural transformation, necessitating time and investment in training IT staff to proficiently navigate the new platform.

6.4.2.2 Higher Initial Investment in HCI

Although Nutanix has the potential to lower operational expenses in the long run, it often necessitates a substantial initial investment for the implementation of hyper-converged infrastructure. This aspect can be particularly challenging for smaller organizations with limited financial resources. Nevertheless, the initial capital outlay is generally balanced by the subsequent decrease in operational costs, especially as companies grow. For instance, a mid-sized manufacturing company may experience surprise at the costs associated with migrating to Nutanix. They must evaluate how the upfront investment can be recovered through ongoing savings in hardware, management, and overall operational efficiency.

6.4.2.3 Vendor Lock-In Concerns

Nutanix minimizes reliance on hardware vendors by enabling the use of standard commodity hardware; however, some organizations might find themselves constrained within Nutanix's software ecosystem. Although Nutanix facilitates multi-cloud strategies, transitioning to an alternative platform after committing to Nutanix's stack can be intricate and expensive. For instance, consider a telecommunications company that depends entirely on Nutanix for its infrastructure. If this company later opts to switch to a different system, the migration process could prove to be more complex and costly than expected, especially given the extensive integration that Nutanix promotes across various cloud environments.

6.4.2.4 Limited Ecosystem Compared to VMware

VMware has established itself as a leader in the virtualization sector for many years, creating an extensive network of partners, third-party integrations, and supplementary services. In contrast, Nutanix, while experiencing significant growth, may have a comparatively smaller ecosystem. Companies that depend on particular VMware integrations might discover that Nutanix's ecosystem does not fully meet their needs. For instance, a global retail chain could depend on third-party applications that are closely integrated with VMware. Although Nutanix provides many similar functionalities, there could be compatibility challenges with certain specialized applications, potentially resulting in extra integration efforts or expenses.

6.4.2.5 Perceived Risk of Transitioning from VMware

For organizations heavily invested in VMware, transitioning to Nutanix may appear to be a daunting prospect. VMware has established a strong reputation for reliability, and any shift to a new platform inherently involves the risk of unexpected obstacles. Stakeholders might be reluctant to make the change, especially if they perceive Nutanix as a relatively new entrant in the industry.

Take, for instance, a global banking institution that relies on VMware for its essential operations. While Nutanix could provide enhanced performance and cost savings, the perceived risks associated with migrating sensitive information and critical workloads may lead decision-makers to hesitate, concerned about possible disruptions during the transition.

Embrarking on Nutanix offers numerous benefits, including streamlined infrastructure management, cost reductions, and improved security. Its ability to scale and integrate seamlessly with cloud services makes it an attractive choice for organizations aiming for agility in a complex IT environment. Nevertheless, decision-makers must also consider potential drawbacks, such as the initial learning curve, upfront investment, and worries about vendor dependency.In the end, the choice to implement Nutanix depends on the unique requirements and objectives of the organization. For those looking to move away from traditional, cumbersome infrastructures while advancing their digital transformation efforts, Nutanix presents a strong case. However, conducting a thorough cost-benefit analysis and developing a strategic migration plan are crucial for ensuring a successful transition.

6.4.3 Takeaway and Summary from the Author's Point of View

As both an author and a technical strategist, I contend that grasping the trade-offs and opportunities offered by Nutanix extends beyond merely choosing another software solution; it involves strategically positioning your organization for sustainable growth and resilience amid the fast-paced landscape of digital transformation. The choice to implement Nutanix instead of a well-established competitor like VMware is complex, influenced by numerous technical, financial, and strategic factors.

6.4.3.1 The Power of Simplified Management

One of the primary insights from this chapter is the effectiveness of simplified management. As companies expand, the intricacies of overseeing their IT infrastructure also increase. Historically, organizations have had to manage various separate systems for computing, storage, and networking. Each of these systems necessitates specialized teams, which adds layers of complexity that hinder innovation and delay response times. Nutanix revolutionizes this approach. By integrating these functions into a unified, software-defined platform, Nutanix enables organizations to oversee their entire infrastructure from a single, user-friendly console.

This integration not only enhances operational efficiency but also transforms how organizations allocate their resources. Previously, substantial time, effort, and financial resources were dedicated to maintaining systems and addressing urgent issues. With Nutanix's streamlined methodology, IT teams can redirect their attention from routine maintenance to strategic initiatives that enhance business value. This transition represents more than just an operational adjustment; it signifies a shift in the perception of IT within the organization. Rather than being viewed merely as a cost center, IT can evolve into a catalyst for innovation and a competitive edge.

6.4.3.2 The Financial Equation

Nutanix presents a compelling value proposition from a cost perspective. While the initial investment in hyper-converged infrastructure (HCI) may appear high, the long-term savings quickly compensate for this expense. Nutanix minimizes the reliance on expensive specialized hardware, reduces the physical space required for data centers, and decreases energy and cooling expenses. Furthermore, Nutanix's predictable scaling model enables businesses to expand their infrastructure gradually, preventing over-provisioning and unnecessary costs for unused resources.

For instance, consider a company preparing for significant growth. With conventional infrastructure, this organization would likely need to make considerable hardware investments, much of which could remain underutilized during off-peak periods. Nutanix provides a more adaptable solution, allowing companies to adjust their infrastructure in response to actual demand, paying solely for what they consume. This transition from capital expenditures to operational expenditures helps organizations better align their IT expenses with their growth trajectory.

6.4.3.3 The Cloud Dilemma: Adaptability and Cohesion

In today's cloud-driven world, flexibility is king. One of Nutanix's standout features is its ability to integrate seamlessly with public cloud environments like AWS, Microsoft Azure, and Google Cloud. This capability gives businesses the flexibility to deploy a hybrid cloud strategy, balancing on-premises workloads with cloud-based solutions. The ease with which Nutanix enables cloud integration is a game-changer for organizations that need to quickly scale resources or balance workloads across different environments. For example, imagine an online retailer facing a surge in traffic during a holiday shopping season. In a traditional infrastructure model, they might have to invest in expensive hardware upgrades to handle the increased demand. With Nutanix's hybrid cloud capabilities, they can quickly scale resources in the cloud, meet customer demand, and scale back down when traffic normalizes, all without costly investments in hardware.

The Nutanix Cloud Platform (NCP) and Nutanix Cloud Clusters (NC2) empower organizations to effortlessly extend their infrastructure into the public cloud, thereby improving operational agility and cost-effectiveness. NC2 facilitates hybrid and multi-cloud setups, enabling businesses to deploy their applications across both private and public clouds while ensuring a consistent management experience. This capability allows companies to scale their operations as needed, featuring intelligent workload placement that supports cloud bursting for temporary infrastructure requirements or during peak demand periods.

By partnering with major cloud providers such as AWS and Microsoft Azure, Nutanix delivers a comprehensive solution for organizations seeking geographic redundancy, disaster recovery, and the swift migration or modernization of applications without extensive refactoring. This strategy simplifies the management of diverse cloud environments, providing a cohesive and economical solution for hybrid cloud operations.

6.4.3.4 Security by Design

Security has evolved from being a secondary consideration to a critical necessity for any IT infrastructure, particularly as threats grow increasingly complex. Nutanix adopts a security-first strategy, incorporating features such as micro-segmentation, zero-trust architecture, and integrated disaster recovery. For instance, Nutanix Flow's micro-segmentation effectively isolates workloads, thereby restricting the movement of potential threats within the network. This approach significantly limits an attacker's ability to navigate laterally within the system in the event of a breach. For organizations that handle sensitive customer information, including healthcare providers and financial institutions, these inherent security features offer substantial reassurance. Nutanix guarantees that data is protected at every level of the infrastructure, from the hypervisor to the network, thereby minimizing the risk of breaches that could result in reputational harm and financial loss.

6.4.3.5 Scalability and Flexibility for Growing Businesses

A key insight from this analysis is Nutanix's capability to grow alongside business expansion. Companies are no longer required to invest significantly in hardware and software upfront to meet future demands. Nutanix enables them to begin with a modest setup and gradually increase their resources by adding nodes and capacity as necessary, all without the downtime or disruptions commonly associated with traditional infrastructure upgrades. This flexibility is especially advantageous for sectors experiencing rapid growth or those with variable demands, such as e-commerce, financial services, and media and entertainment. For instance, a media company that is expanding and producing high-demand video content can swiftly enhance its storage and computing capabilities as its audience increases, alleviating concerns about capacity shortages or excessive investment in infrastructure that may remain unused during off-peak periods.

6.4.3.6 Challenges and Considerations

Transitioning to Nutanix presents its own set of challenges, as is the case with any solution. A significant factor for organizations making this shift is the initial learning curve. Although Nutanix ultimately streamlines infrastructure management, IT teams may need specialized training to fully utilize the platform's features. Thus, adopting Nutanix signifies both a technological and cultural transformation within the IT department. Furthermore, while Nutanix can lead to cost savings in the long run, the initial investment in hyper-converged infrastructure may pose a challenge for smaller organizations with constrained budgets. Decision-makers must carefully consider the long-term advantages of Nutanix's lower operational costs in relation to the upfront capital required for implementation. This challenge can be addressed through strategic planning, ensuring that infrastructure investments align with business objectives, and investigating financing options that allow for cost distribution over time.

6.4.3.7 Vendor Lock-In Concerns

Nutanix provides a degree of flexibility by enabling the use of various cloud environments and minimizing dependence on proprietary hardware. However, there remains a potential risk of vendor lock-in at the software level. Once an organization implements Nutanix's HCI platform, transitioning to an alternative system later on could be challenging and expensive, especially if Nutanix has been extensively integrated into the organization's processes. It is essential for decision-makers to consider this risk in relation to the advantages of adopting Nutanix while also reflecting on the organization's long-term IT strategy.

6.4.3.8 Final Thoughts

As the author of this chapter, my objective is to deliver a comprehensive, objective, and insightful evaluation of Nutanix as an alternative to VMware. The benefits are clear – streamlined management, cost efficiency, scalability, seamless cloud integration, and improved security. Nevertheless, transitioning to Nutanix is not universally applicable. The specific requirements of each organization, along with industry standards and future aspirations, will determine if Nutanix is the appropriate choice. For organizations aiming to move away from the intricacies of traditional infrastructure and adopt hyper-converged, cloud-connected IT solutions, Nutanix presents a strong case. However, like any significant IT decision, achieving success requires meticulous planning, detailed assessment, and a solid grasp of both the potential benefits and associated risks.

In summary, Nutanix's capability to consolidate and simplify infrastructure management, along with its robust security features and cloud-native adaptability, positions it as a formidable option for organizations seeking to enhance their IT operations. Nonetheless, decision-makers must navigate this transition with clarity, fully aware of the initial costs, training requirements, and long-term obligations involved. When implemented with a strategic approach, Nutanix can facilitate new levels of efficiency and innovation, equipping companies for success in an ever-evolving digital environment.

6.5 Nutanix in Multi-cloud and Hybrid Cloud Environments

In the current digital landscape, organizations are no longer limited to traditional data center frameworks. The rise of cloud computing has transformed business operations, providing unprecedented levels of agility, scalability, and efficiency. As the cloud ecosystem continues to develop, the focus has shifted from selecting a single cloud platform to managing a diverse array of environments where data, applications, and services are distributed across various clouds – whether public, private, or a hybrid of both. This is where Nutanix excels, offering businesses the tools to effectively manage the intricacies of multi-cloud and hybrid cloud settings.

Nutanix's architecture, designed to function seamlessly within both multi-cloud and hybrid cloud frameworks, empowers organizations to implement a cloud strategy tailored to their specific operational requirements. The ability to select an optimal combination of on-premises infrastructure, public cloud, and private cloud represents a significant advantage for businesses navigating today's rapidly evolving market.

6.5.1 The Multi-cloud Challenge

Consider this scenario: A worldwide e-commerce enterprise has experienced remarkable growth. Their operations extend across various continents, leading to an exponential increase in infrastructure demands. To address this rising need, they have implemented a multi-cloud strategy. They utilize AWS for customer databases, Microsoft Azure for analytics, and Google Cloud for machine learning, while also maintaining a private cloud for sensitive customer information that must adhere to stringent regulatory standards.

Although this strategy offers significant flexibility, it presents numerous challenges. The management of these varied environments often involves navigating different tools, interfaces, billing systems, security measures, and compliance requirements. This complexity can result in operational inefficiencies, increased costs, and the potential for data silos, complicating the IT teams' ability to achieve a comprehensive view of their entire infrastructure.

Now envision this same enterprise partnering with Nutanix to streamline and enhance their multi-cloud strategy. Nutanix's platform allows them to oversee all these environments – AWS, Azure, Google Cloud, and their private cloud – from a unified interface. By offering centralized management, Nutanix assists IT teams in minimizing complexity, improving control over their infrastructure, and optimizing resources across various cloud providers.

6.5.2 Nutanix: A True Enabler of Hybrid Cloud

The hybrid cloud represents a strategic approach adopted by organizations seeking to integrate on-premises infrastructure with cloud-based solutions. Consider a large financial services firm as an example. This institution maintains its most sensitive customer data on-site to comply with strict regulatory standards while simultaneously aiming to utilize the public cloud for its innovation labs, artificial intelligence projects, and mobile application development. The hybrid cloud framework allows them to safeguard critical workloads on-premises while taking advantage of the flexibility and scalability offered by public cloud services.

Nonetheless, the hybrid cloud model presents its own set of challenges. Many organizations face difficulties in achieving a seamless integration between their on-premises systems and public cloud platforms. This is where Nutanix provides a valuable solution, effectively connecting traditional data centers with public cloud environments.

By utilizing Nutanix Clusters, the financial institution can effortlessly extend its private cloud into the public cloud without needing to modify its applications. This capability allows the firm to quickly provision additional resources in AWS or Azure to address peak demand or fulfill disaster recovery needs, all while maintaining strict adherence to data governance, compliance, and security protocols. Nutanix empowers this organization to function as a unified cloud entity, removing barriers and friction between on-premises and cloud operations.

6.5.3 Unified Management Across Clouds

Nutanix's unified management approach serves as the foundation for its multi-cloud and hybrid cloud capabilities. For instance, a healthcare organization might utilize Microsoft Azure for patient data analytics, AWS for storage and backup solutions, and an on-premises Nutanix setup for everyday hospital functions. Traditionally, overseeing these distinct environments necessitated specialized teams for each cloud service, each equipped with unique tools and management interfaces. This scenario not only complicates operations but also significantly increases management burdens.

Nutanix revolutionizes this process by providing a unified management layer via its Prism console. IT administrators can efficiently oversee all cloud environments – whether public, private, or hybrid – through a single interface. Tasks such as managing virtual machines, monitoring workloads, and enforcing security policies can all be executed cohesively, thereby minimizing complexity and operational challenges.

Additionally, Nutanix's one-click operations facilitate the deployment and scaling of hybrid and multi-cloud environments, eliminating the need to navigate the intricacies of various cloud providers. For healthcare organizations, where timely responses are crucial and downtime can pose serious risks, the ability to streamline operations and maintain continuity across different cloud platforms is essential. Nutanix empowers these organizations to concentrate on their primary objective – delivering exceptional patient care – rather than managing fragmented cloud infrastructures.

6.5.4 Optimizing Costs in a Multi-cloud World

To optimize costs in a multi-cloud environment, Nutanix offers a strong solution through Nutanix Cloud Manager's Cost Governance (formerly Nutanix Beam). This platform provides real-time insights into multi-cloud usage across major public clouds like AWS, Azure, and Google Cloud. It helps businesses identify inefficiencies, manage resources, and reduce unnecessary expenses.

For example, a media company that uses Google Cloud for streaming and AWS for storage can leverage Nutanix Cost Governance to monitor its cloud expenditure. They may discover underutilized storage in AWS or over-provisioned compute resources in Google Cloud, allowing them to optimize their spending accordingly. This enables businesses to make data-driven decisions to reduce costs while maintaining operational efficiency.

Nutanix's cost optimization tools are particularly beneficial for organizations with fluctuating resource needs, such as e-commerce companies preparing for high-demand events like Black Friday. By using Nutanix Cloud Manager, companies can dynamically adjust cloud usage and ensure financial sustainability, making it a valuable asset in multi-cloud management.

For seamless application mobility, Nutanix also allows businesses to easily move applications between on-premises and public clouds without re-architecting, reducing downtime and migration risks. This flexibility is essential for organizations that need to adapt quickly to changing business demands.

6.5.5 Security and Compliance Across Clouds

Operating within a multi-cloud or hybrid cloud framework presents considerable security challenges, particularly when dealing with sensitive information. Organizations in heavily regulated sectors, including finance, healthcare, and government, are required to follow stringent data protection and compliance regulations. Consequently, transitioning to the cloud can appear to be a daunting task.

Nutanix addresses these concerns with its integrated security features, such as Nutanix Flow and data encryption, which safeguard data whether it resides on-premises or in the cloud. For instance, a financial institution can utilize Nutanix to implement micro-segmentation, effectively isolating workloads and managing access to sensitive customer information, all while ensuring adherence to industry regulations. Nutanix Flow facilitates the application of these security measures across both public and private clouds, creating a cohesive security framework that mitigates the risk of data breaches.

Moreover, Nutanix provides tools like Calm for orchestration and automation, which play a crucial role in consistently enforcing security and compliance standards across various environments. This capability is essential for organizations that require stringent oversight of their data and must comply with regulatory mandates, regardless of where their workloads are deployed.

6.5.6 Future-Proofing with Nutanix

In an era where technological advancements occur at an extraordinary rate, ensuring future readiness is a primary focus for IT executives. Nutanix's strategy for multi-cloud and hybrid cloud environments delivers the necessary flexibility and scalability to meet evolving demands. Whether your organization aims to enter new markets, launch innovative digital services, or address the challenges posed by regulatory changes, Nutanix lays the groundwork for growth and innovation.

For instance, a global retailer seeking to expand into new territories can leverage Nutanix to swiftly set up local infrastructure in the public cloud while retaining centralized oversight of operations. As the retailer expands, Nutanix's capability to seamlessly integrate with additional cloud providers or enhance on-premises setups guarantees that they can fulfill demand without compromising on performance or security.

In summary, Nutanix's multi-cloud and hybrid cloud functionalities empower organizations to operate with agility, manage complexities, optimize expenses, and maintain security across various environments. By offering a cohesive platform that connects on-premises infrastructure with public clouds, Nutanix enables businesses to fully harness the advantages of the cloud without being restricted to a single vendor or technology stack. This adaptability, along with Nutanix's user-friendly management, cost-saving, and security features, positions it as a transformative solution for companies aiming to excel in the multi-cloud landscape. Whether your organization is embarking on its cloud journey or already navigating a hybrid environment, Nutanix provides the essential tools to streamline operations, foster innovation, and ensure your IT strategy is future-ready.

6.6 Case Study: Nutanix – A Challenger's Approach

Let's consider, for instance, Xtech, a prominent player in global manufacturing, is currently facing challenges related to the growing complexity and costs of its extensive VMware-based infrastructure. The IT teams at Xtech are encountering significant issues, such as rising operational costs, frequent downtimes during system upgrades, and challenges in scaling to accommodate the company's rapid expansion. Despite relying on VMware as their virtualization platform for many years, it has become clear that its traditional model no longer aligns with Xtech's changing business goals and future scalability requirements.

Recognizing the need for change, the executive leadership, including the CIO, decided to investigate alternative solutions. With a focus on cost-effectiveness, scalability, and streamlined management, they began to explore hyper-converged infrastructure (HCI) options. After a thorough assessment of various solutions, Nutanix emerged as a strong candidate to replace their VMware setup. This case study delves into Xtech's transformative journey as they opted for Nutanix's innovative approach over VMware, ultimately reshaping their IT environment and ensuring their operations are future-ready.

6.6.1 The Challenge

Xtech faced a range of challenges, and the choice to move away from VMware was not taken lightly. Over time, their VMware setup had become unwieldy, leading to increased expenses associated with licensing, hardware dependencies, and growing maintenance demands. The process of scaling to accommodate new production requirements was labor-intensive and often necessitated considerable manual effort. Moreover, frequent system outages during updates and patches adversely affected essential manufacturing processes, resulting in expensive production delays.

The intricacies of managing a hybrid cloud environment – where some operations had to stay on-premises while others transitioned to the public cloud – added to their difficulties. VMware's restricted flexibility in managing hybrid clouds created barriers between on-premises and cloud operations, resulting in a lack of cohesive visibility and control over their infrastructure.

The CIO, responsible for lowering operational costs while enhancing agility, began exploring solutions that would provide improved control over hybrid environments, simplify management, and boost scalability – all while maintaining performance and security.

6.6.2 The Nutanix Solution

Nutanix emerged as a formidable competitor in the HCI sector, committed to addressing these challenges directly. The company introduced a platform that consolidated Xtech's infrastructure, providing effortless scalability, reduced operational expenses, and improved performance within hybrid cloud settings. By integrating software-defined storage, networking, and computing, Nutanix offered a solution designed to eradicate numerous inefficiencies associated with the traditional VMware system.

- **Simplicity through unified management:** Nutanix provided a streamlined solution with Prism, a unified management interface that consolidated all workloads, whether on-premises or in the cloud, unlike VMware, which necessitated the use of multiple management consoles. This integration significantly eased daily operations, enabling IT personnel to oversee Xtech's complete infrastructure from a single platform. With Prism's integrated automation and AI-powered analytics, Xtech was able to swiftly pinpoint bottlenecks, forecast future resource requirements, and automate routine processes such as system updates and patch management – eliminating the downtime that had been a persistent issue in their VMware setup.

- **Cost-effectiveness:** Nutanix's comprehensive software pricing structure attracted the attention of Xtech's management team. In contrast to VMware, which imposed extra licensing fees for individual features, Nutanix offered a unified software stack that encompassed a wide range of functionalities, including data deduplication, compression, disaster recovery, and hybrid cloud management tools. This approach resulted in a notably lower total cost of ownership (TCO) over time. The company projected a 30% decrease in operational expenses within the first year of deployment, primarily by removing numerous licensing, hardware, and maintenance costs associated with VMware.

- **Scalability on demand:** Nutanix's hyper-converged architecture enabled Xtech to scale resources with ease. In contrast, scaling with VMware typically required the acquisition of extra hardware and intricate licensing for each element, accompanied by protracted setup periods. Nutanix simplified this process by allowing the addition of nodes to the cluster, which seamlessly integrated into the current infrastructure. As a result, Xtech was able to increase capacity in response to rising demand without experiencing significant downtime or complicated reconfigurations.

- **Migration strategy:** Xtech employed a replatforming strategy during its migration to Nutanix, capitalizing on Nutanix's in-built features while ensuring business continuity. The IT team worked closely with Nutanix's migration experts, leveraging Nutanix Move, a dedicated migration tool designed to streamline VM migrations from VMware ESXi to Nutanix AHV. Nutanix Move supports features like automated VM conversion, built-in replication, and real-time synchronization, which minimized disruptions during the transition. To avoid any operational downtime, they established a parallel Nutanix environment mirroring the existing VMware setup. The migration was executed in phases, starting with non-critical workloads. Over time, the company migrated core business applications, addressing any challenges encountered with the support of Nutanix's expert team, before fully decommissioning the VMware environment. This approach allowed Xtech to maintain daily operations seamlessly while transitioning to Nutanix's platform, with tools like Nutanix Move ensuring minimal downtime and simplified network mapping throughout the migration process.By incorporating Nutanix Move, a native cross-hypervisor migration solution, Xtech benefited from a smoother transition and reduced risks compared to manual methods.

- **Seamless hybrid cloud integration:** Nutanix's inherent hybrid cloud features significantly influenced Xtech's choice to transition. The platform enabled them to effortlessly integrate their on-premises infrastructure with the public cloud, resulting in a cohesive hybrid setup. By leveraging Nutanix Clusters on AWS, Xtech successfully migrated their on-premises workloads to the cloud without the necessity of refactoring applications or altering network configurations. This development was particularly transformative for Xtech, as it allowed them to keep specific sensitive manufacturing processes on-premises while taking advantage of the cloud's scalability and flexibility for other operations.

6.6.3 The Results

1. **Reduced complexity and improved efficiency:** Xtech saw a swift enhancement in its operational efficiency. The integrated management console significantly decreased the time needed for routine tasks, allowing IT resources to concentrate on innovation instead of maintenance. The team noted a 50% decrease in system management overhead within the initial months, and the automated patching system effectively eradicated unplanned downtime.

2. **Lower total cost of ownership:** In the initial year, Xtech experienced a 30% decrease in expenses, largely attributed to Nutanix's streamlined licensing approach and diminished reliance on hardware. The capacity to scale effortlessly without the necessity of over-purchasing hardware in advance allowed Xtech to pay solely for the resources they required, precisely when they required them.

3. **Enhanced business agility:** Through Nutanix's hybrid cloud solutions, Xtech successfully adapted to emerging business opportunities with agility. Whether introducing a new product line or entering a new market, the IT infrastructure was capable of scaling in real time to support the business's requirements. The flexible licensing and seamless scalability across both on-premises and public cloud environments enabled Xtech to respond to demand fluctuations without the delays that were once typical with VMware.

6.6.3.1 Conclusion: A Strategic Pivot for Growth

By moving from VMware to Nutanix, Xtech not only achieved cost savings and reduced operational complexity but also gained enhanced agility and innovation capabilities. The IT team, previously overwhelmed by the demands of infrastructure management, could now concentrate on delivering business value through technology. The industry acknowledged that Nutanix played a crucial role in transforming Xtech's IT infrastructure from a mere cost center into a valuable strategic asset.

For companies like Xtech, Nutanix is not merely an alternative to VMware; it signifies a strategic shift toward a future where simplicity, flexibility, and scalability dominate IT decision-making. This transition enabled Xtech to innovate more rapidly, scale effectively, and attain operational excellence, positioning them for sustained growth in a cloud-centric environment. This case study highlights the importance of considering "challenger" solutions as viable options for progressive, growth-focused organizations.

6.7 Summary

In this chapter, we examined the significant advantages and obstacles associated with the adoption of Nutanix, a prominent player in hyper-converged infrastructure (HCI). Nutanix streamlines IT operations by merging storage, computing, and networking into a cohesive platform, which enhances scalability and simplifies management. Its capacity to minimize operational complexity, reduce expenses, and improve performance makes it an attractive option for organizations in search of contemporary IT solutions. We also analyzed Nutanix's proficiency in hybrid and multi-cloud settings, highlighting its seamless integration with public cloud services such as AWS, Azure, and Google Cloud. Key features like automation, one-click upgrades, and integrated disaster recovery capabilities are essential for boosting operational efficiency, particularly for organizations focused on minimizing downtime and complexity. Nonetheless, the shift to Nutanix presents certain challenges. The initial investment and the learning curve related to hyper-converged infrastructure can be significant barriers, particularly for businesses with intricate legacy systems. Adopting Nutanix necessitates a thorough evaluation of current infrastructure and long-term objectives, as well as ensuring that IT teams are adequately prepared for the transition. In summary, Nutanix offers a compelling alternative to conventional IT frameworks, delivering flexibility, efficiency, and scalability. This chapter has provided decision-makers with a comprehensive understanding of Nutanix's advantages and potential drawbacks, enabling them to make well-informed choices regarding their infrastructure modernization efforts.

CHAPTER 7

Proxmox, RedHat, and Beyond

As organizations worldwide adapt to the increasing demands of the digital landscape, the urgency to upgrade IT infrastructure has grown significantly. VMware has long been a leader in the virtualization sector, known for its enterprise-grade reliability, features, and support that have made it indispensable for numerous businesses. However, the technological environment is evolving, along with the requirements of contemporary enterprises. More organizations are now seeking flexible, cost-efficient, and innovative alternatives that not only satisfy current needs but also enable future growth and adaptability. This is where Proxmox and RedHat come into play, emerging as notable contenders in the virtualization arena, each providing a viable alternative to VMware's conventional model.

Proxmox and RedHat have gained popularity for their capacity to meet the demands of organizations looking for more customizable and scalable options. These platforms offer advantages that VMware may find challenging to deliver in today's fast-paced environment, such as freedom from restrictive licensing agreements, enhanced flexibility for hybrid and multi-cloud configurations, and access to dynamic open-source communities. However, transitioning away from VMware, a trusted industry leader for many years, is a significant undertaking. Decision-

© Sumit Bhatia and Chetan Gabhane 2025
S. Bhatia and C. Gabhane, *Navigating VMware Turmoil in the Broadcom Era*,
https://doi.org/10.1007/979-8-8688-1264-4_7

makers, particularly CXOs, must thoroughly comprehend the ramifications of such a change. What impact will this shift have on operational efficiency? How will it influence costs both in the short and long term? Will these alternatives fulfill the organization's security and compliance requirements, or will they introduce new obstacles?

This chapter aims to guide those contemplating Proxmox, RedHat, or similar platforms as potential substitutes for VMware. By examining these alternatives in depth, we seek to clarify the unique value propositions of each platform and how they tackle modern infrastructure challenges. IT leaders are increasingly confronted with decisions that affect not only their technical teams but the broader business ecosystem. The essential factor lies in understanding these dynamics.

7.1 Evaluating Alternative Offerings and Architectures

Let's evaluate alternative Proxmox and RedHat for VMware in this chapter. We also evaluate the architecture of each on a high level for a fair comparison.

7.1.1 Proxmox: Flexibility Through Simplicity

Proxmox, an open-source solution, has become a practical choice for organizations aiming to lessen their dependence on expensive proprietary virtualization options. Centered around the KVM (kernel-based virtual machine) hypervisor, Proxmox provides an efficient and streamlined virtualization experience, merging robust features with user-friendliness. A key benefit is its capability to establish and oversee a comprehensive virtualized infrastructure without the constraints of costly licensing agreements that often burden traditional providers like VMware. Please refer to Figure 7-1, which showcases user tool option availability to manage the Proxmox platform in a controlled way.

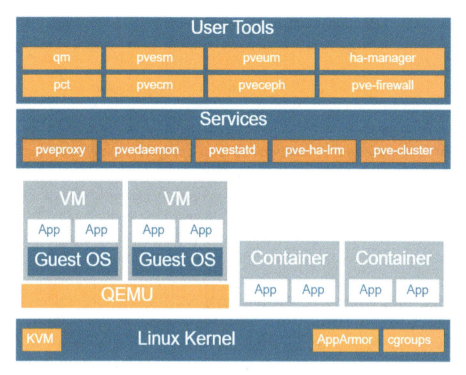

Figure 7-1. *Proxmox architecture on a high level*

In Proxmox's robust ecosystem, each tool and service plays a unique role in ensuring seamless operation. **qm** manages virtual machines, allowing users to create, configure, and control VMs. **pvesm** handles storage, organizing VM disks, containers, and ISO images in dedicated pools, while **pveum** manages user access, setting permissions to keep the environment secure. For clustering, **pvecm** joins nodes together, enabling high availability, which **ha-manager** orchestrates by monitoring and restarting VMs if a node fails. **pveceph** integrates Ceph storage, providing scalable and distributed storage solutions, and **pct** focuses on container management. **pvc-firewall** safeguards the environment with customizable security rules. Key services keep Proxmox responsive: **pveproxy** for the web interface, **pvedaemon** for API requests, **pvestatd** for resource monitoring, **pve-ha-lrm** for local HA resource management, and **pve-cluster** for syncing configurations across nodes. Together, these components create a flexible and reliable proxmox virtualization platform.

For businesses that value adaptability and budget management, Proxmox stands out as an attractive alternative. Its lightweight design enables IT teams to swiftly deploy virtual machines, manage intricate clusters, and leverage containers – all from a single, cohesive platform. The user-friendly management interface of Proxmox eliminates the steep learning curve typically associated with proprietary systems, making it especially beneficial for smaller organizations or those with limited IT capabilities. This accessibility allows teams to concentrate on their primary applications without being hindered by unnecessary complexities.

A notable instance can be observed in mid-sized manufacturing firms, where operational budgets are frequently constrained. Proxmox provides these companies with a chance to optimize resources without compromising performance. Instead of incurring substantial licensing costs, organizations can allocate their budgets toward innovation and expansion. Additionally, Proxmox's compatibility with both virtual machines and containers enhances flexibility, enabling IT departments to efficiently manage diverse workloads. In sectors where rapid deployment and cost efficiency are critical, Proxmox offers a distinct advantage over more rigid and expensive alternatives.

7.1.2 RedHat: The Hybrid Cloud Innovator

RedHat has established itself as a frontrunner in hybrid cloud solutions, particularly with its OpenShift platform, which effectively incorporates Kubernetes and containerization into its framework. For organizations deeply engaged in cloud-native applications, RedHat offers the scalability and innovative capabilities essential for thriving in a digital-centric landscape. OpenShift is specifically tailored to manage containerized workloads across both private and public clouds, enabling businesses to develop and deploy applications with a level of agility that traditional virtualization platforms, such as VMware, may find challenging to achieve. Figure 7-2 shows how customer-executed application portability can be achieved on the RedHat platform using OpenShift.

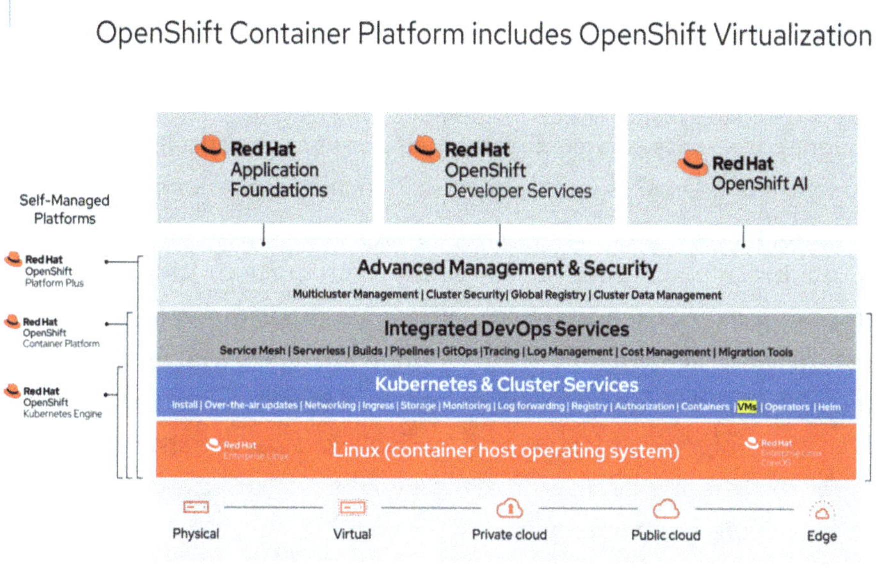

Figure 7-2. RedHat platform with OpenShift

The OpenShift Container Platform from RedHat represents a comprehensive ecosystem tailored to facilitate a wide range of enterprise applications across various deployment settings, including on-premises, cloud, and edge environments. This architecture incorporates multiple service layers, allowing organizations to efficiently build, deploy, and manage containerized applications. At its core is RedHat Enterprise Linux (RHEL), specifically the RedHat Enterprise Linux CoreOS, which serves as the operating system for container hosting. This foundational layer provides a consistent, secure, and optimized environment for container orchestration, effectively connecting physical, virtual, private, and public cloud infrastructures, as well as edge computing.

Above this foundational operating system is the Kubernetes and Cluster Services layer. Kubernetes is central to OpenShift's orchestration capabilities, overseeing clusters across various environments. This layer is equipped with vital services for container management, such as over-the-air updates, networking, ingress, storage, monitoring, log forwarding, and registry services. It also integrates tools like virtual machines, operators, and Helm to enhance control and automation in deployment and lifecycle management.Further up, we find the Integrated DevOps Services layer, which serves as a robust framework for contemporary DevOps methodologies. It includes tools for continuous integration and continuous deployment (CI/CD), pipelines, and GitOps, facilitating smooth transitions from code to production. Additionally, it encompasses critical operational components such as tracing, logging, and cost management, providing development teams with the insights and resources necessary to optimize performance and manage resources effectively. Migration tools are also available to assist in transitioning from legacy systems or other platforms to OpenShift, ensuring a seamless shift and continuity in operations.

The advanced management and security layer offers improved oversight and protection features. It encompasses multi-cluster management tailored for extensive, distributed environments; strong cluster security to protect data and workloads; a global registry; and extensive cluster data management functionalities. This layer exemplifies OpenShift's dedication to resilience and security, allowing teams to oversee multiple clusters from a unified control plane while maintaining compliance and data integrity across different environments. At the pinnacle of this architecture, the RedHat Application Foundations and RedHat OpenShift Developer Services layers facilitate application development. Application Foundations delivers critical middleware, runtime, and data solutions for rapid application building and deployment, while Developer Services equips developers with essential tools for efficient work, including IDE integrations, command-line utilities, and managed services.

Moreover, OpenShift's AI capabilities, represented by RedHat OpenShift AI, integrate machine learning and artificial intelligence into the platform, enabling organizations to utilize AI/ML workloads alongside conventional applications. This integration facilitates the incorporation of AI-driven insights and automation into business applications, thereby enhancing decision-making and operational efficiency.

Collectively, these layers form a comprehensive and cohesive platform that meets the needs of contemporary enterprises, promoting scalability, security, and flexibility across various infrastructure environments. The OpenShift Container Platform delivers a full-stack, end-to-end solution that caters to the requirements of both development and operations teams, positioning it as a formidable option for enterprises seeking to expedite their digital transformation initiatives.

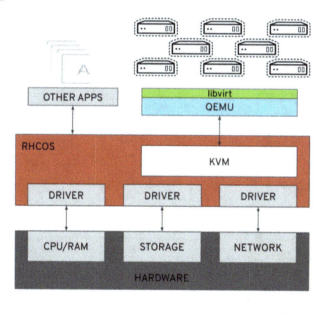

Figure 7-3. *KVM hypervisor for OpenShift*

Figure 7-3 shows OpenShift Virtualization utilizes KVM (kernel-based virtual machine), a well-established hypervisor within the Linux kernel that has been in production for over 10 years. It is a trusted component across various RedHat offerings, including RedHat Virtualization, OpenStack Platform, and RedHat Enterprise Linux (RHEL). KVM acts as a fundamental virtualization element, providing high performance and security for running virtual machines (VMs). The execution of VMs is supported by QEMU, while libvirt serves as an abstraction layer, streamlining VM management through a unified API. This architecture supports x86 bare metal environments, ensuring effective virtualization on physical hardware, with RedHat Enterprise Linux CoreOS (RHCOS) functioning as the host operating system, adeptly managing resources such as CPU, memory, storage, and networking through its core drivers.

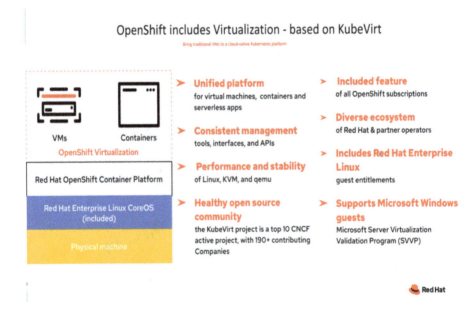

Figure 7-4. *OpenShift virtualization based on KubeVirt*

OpenShift Virtualization is shown in Figure 7-4 as an enhancement of the OpenShift Container Platform built on KubeVirt, which integrates virtual machines (VMs), containers, and serverless applications into a unified platform. This integration allows organizations to effectively manage both traditional and cloud-native workloads using consistent management tools, interfaces, and APIs. Leveraging the power of Linux, KVM, and QEMU, OpenShift Virtualization delivers high performance and stability, which are essential for enterprise-level applications.The platform's capabilities are further strengthened by its open-source nature, with KubeVirt recognized as one of the leading projects under the Cloud Native Computing Foundation (CNCF), supported by contributions from over 190 companies that drive innovation and community support. OpenShift Virtualization is available with all OpenShift subscriptions, making it readily accessible for current users. It features a broad ecosystem, backed by RedHat and partner operators, to meet various deployment requirements. Additionally, RedHat Enterprise Linux guest entitlements are included, providing added value for Linux users, while support for Microsoft Windows guests is available through the Microsoft Server Virtualization Validation Program (SVVP), enhancing compatibility. Overall, OpenShift Virtualization offers a robust, integrated solution for hybrid environments, enabling enterprises to seamlessly transition between traditional and modern application architectures.

7.1.3 Balancing Innovation with Reliability

For decision-makers, the selection between Proxmox and RedHat is not simply a matter of one being better than the other. It involves finding the solution that best aligns with the specific objectives and operational needs of the organization. Proxmox stands out for its emphasis on cost efficiency, ease of use, and adaptability, making it an excellent choice for organizations aiming to optimize their IT expenditures while retaining essential capabilities. In contrast, RedHat's OpenShift is particularly advantageous in scenarios where scalability, cloud-native development, and hybrid cloud infrastructures are paramount.

A crucial aspect of assessing these platforms is grasping the organization's long-term aspirations. If a company intends to expand its cloud operations or implement DevOps methodologies, RedHat's integration with Kubernetes provides a robust, future-ready option. Conversely, for organizations focused on minimizing costs while retaining oversight of their virtualization setup, Proxmox offers a streamlined, user-friendly alternative that does not necessitate a large initial investment.

Ultimately, the choice to transition from VMware to another platform should be influenced by both present operational demands and future business goals. Each platform presents unique value propositions tailored to different market segments. Proxmox's straightforwardness and cost-effectiveness position it as a compelling option for organizations seeking flexibility without the burden of proprietary licensing fees. On the other hand, RedHat's dedication to hybrid cloud advancements and containerization makes it the preferred option for enterprises managing intricate, cloud-centric workloads.

As businesses continue to adapt to the challenges of digital transformation, the demand for agile and cost-efficient IT solutions is increasingly critical. The decision between Proxmox, RedHat, or any other VMware alternative relies on a thorough assessment of organizational objectives, infrastructure needs, and growth strategies.

7.2 Cost, Management, and Suitability Comparison with VMware

Proxmox and RedHat present significantly more economical options while still maintaining essential features necessary for efficient virtualization and cloud deployment. Proxmox, an open-source platform utilizing the KVM (kernel-based virtual machine) hypervisor, is especially appealing for organizations aiming to optimize their IT expenditures. By adopting Proxmox, companies can avoid expensive licensing fees, allowing them to redirect those funds toward innovation, product development, or enhancing their IT infrastructure. This makes Proxmox an attractive option for small to mid-sized businesses or startups that need strong virtualization capabilities but face budgetary limitations.

RedHat, on the other hand, presents a unique blend of affordability and enterprise-grade functionality, particularly in the hybrid cloud space. RedHat's OpenShift, built on Kubernetes, allows businesses to manage containerized workloads across both private and public clouds with ease. This hybrid cloud approach is invaluable for businesses that are growing quickly and need to maintain flexibility across diverse environments. While RedHat's costs are typically higher than Proxmox's due to its subscription-based model, the value it provides in hybrid cloud environments and container orchestration can offset the initial investment.

For large organizations that operate complex, mission-critical applications across various cloud environments, RedHat provides a distinctive capability to integrate seamlessly with existing infrastructures while ensuring high standards of security and compliance. For instance, a global financial institution handling highly sensitive information may find RedHat's offerings particularly appealing due to its compliance certifications and strong support for hybrid cloud frameworks. In such scenarios, the elevated costs are warranted by the ability to satisfy rigorous regulatory demands while embracing contemporary cloud-native methodologies.

Management complexity represents another significant challenge where VMware's conventional approach may introduce difficulties. VMware environments typically necessitate a team of highly specialized administrators and consultants, resulting in increased staffing expenses and operational intricacies. The implementation of advanced features, updates, and patches often requires downtime, contributing to additional disruptions. While VMware environments are robust, they can also appear inflexible and less responsive, making daily management a resource-intensive process.

In contrast, Proxmox is recognized for its user-friendly management and intuitive interface. Its web-based management console offers a clear, unified perspective of both virtual machines and containers, making it accessible even to teams lacking extensive virtualization knowledge. By streamlining management tasks, Proxmox not only alleviates operational complexity but also reduces the likelihood of downtime during updates or maintenance, allowing businesses to operate smoothly without interruptions. Its integrated support for both virtualization and containerization minimizes the need for multiple tools to manage diverse workloads, further decreasing overhead.

RedHat provides a level of operational flexibility that VMware frequently lacks, especially in the management of hybrid cloud environments. Through OpenShift, organizations acquire a robust platform that combines Kubernetes with DevOps automation, facilitating the scaling of applications across various environments without the need for manual intervention. The platform benefits from strong community support, along with RedHat's dedicated enterprise support options, which offer reassurance and a degree of control that is often missing in more inflexible proprietary solutions.

The suitability for diverse use cases is also crucial in selecting the appropriate platform. For organizations that demand high levels of customization and require detailed control over their IT infrastructure, Proxmox delivers unmatched flexibility. Its open-source framework allows businesses to adapt the platform to their specific requirements, whether that involves integrating with legacy systems or developing highly specialized configurations. Proxmox is particularly well-suited for those in search of a versatile platform without the financial burden associated with proprietary software.

Conversely, for organizations that are transitioning to cloud-native architectures and hybrid cloud environments, RedHat's OpenShift emerges as the premier solution. Its comprehensive integration with Kubernetes positions it as the preferred platform for enterprises aiming to implement containerized applications on a large scale. Although it may not provide the same cost benefits as Proxmox, RedHat's long-term value is evident in its capacity to connect on-premises infrastructure with the public cloud, ensuring smooth operations across both environments.

Table 7-1. *RedHat and Proxmox in comparison with VMware*

Feature	VMware	Proxmox	RedHat (OpenShift)
Cost	• High licensing fees, especially for large-scale deployments	• Free and Proxmox Virtual Environment's source code is published under the free software license GNU AGPL, v3, and thus is freely available for download, use, and share. A Proxmox VE subscription is an additional service program that helps IT professionals and businesses keep Proxmox VE deployments up-to-date.	• Subscription-based pricing, more affordable than VMware but higher than Proxmox also OpenShift has a free version (OKD – Origin Kubernetes Distribution) and several commercial versions that include additional support and functionalities.

(*continued*)

Table 7-1. (*continued*)

Feature	VMware	Proxmox	RedHat (OpenShift)
	• Additional costs for add-ons like disaster recovery, security	• No added costs for basic features	• Costs offset by hybrid cloud and container management capabilities
	• TCO includes specialized skills, updates, and complex integrations	• Low TCO; minimal training required for basic operations	• Higher TCO compared to Proxmox, but competitive in hybrid cloud and compliance
Management	• Complex manage ment requiring specialized teams or consultants	• Simplified management via a unified web interface	• Built-in Kubernetes-based management tools, suitable for containerized environments
	• Strong enterprise-level support and automation tools	• Community-driven support model; easy to deploy and manage	• Strong vendor-backed support with robust automation in hybrid cloud scenarios
	• Requires significant resources for large environments	• Suitable for small businesses with lean IT teams	• Ideal for enterprises needing advanced orchestration and management for hybrid environments

(*continued*)

Table 7-1. (*continued*)

Feature	VMware	Proxmox	RedHat (OpenShift)
Scalability	• Enterprise-grade scalability with robust infrastructure	• Suitable for small to mid-sized deployments	• Highly scalable across multi-cloud environments, with strong support for hybrid cloud setups
	• Higher cost as the infrastructure grows	• Easily scalable without escalating licensing costs	• Scalability through Kubernetes and container orchestration across clouds
Suitability	• ideal for large enterprises and organizations that require a mature and reliable virtualization solution with commercial support.	• Ideal for SMBs or startups looking for cost-effective virtualization	• Best suited for enterprises operating in hybrid cloud or requiring container orchestration
	• Organizations needing a polished experience with comprehensive support	• Cost-conscious businesses aiming for strong virtualization capabilities without vendor lock-in	• Organizations needing compliance, container management, and flexibility across clouds

(continued)

311

Table 7-1. (*continued*)

Feature	VMware	Proxmox	RedHat (OpenShift)
Integration	• Seamless integration with the VMware ecosystem (vSphere, vSAN, etc.)	• Easy integration with existing VMware environments for hybrid setups	• Seamless integration with public clouds like AWS, Azure, and GCP
	• Well-suited for complex environments requiring multi-layered infrastructure	• Allows easy transition from VMware with minimal disruption	• Built for hybrid cloud environments, offering seamless integration across multiple platforms
Security and Compliance	• Enterprise-grade security with various add-ons	• Open-source allows for customizable security, but compliance is user-managed	• Strong compliance certifications, especially for regulated industries
	• Compliance solutions are available but often as added costs	• Suitable for businesses that don't need heavy regulatory oversight	• Perfect for industries like finance, healthcare, and government due to strong security protocols and compliance support

(*continued*)

Table 7-1. (*continued*)

Feature	VMware	Proxmox	RedHat (OpenShift)
Performance	• Best-in-class for enterprise-grade, mission-critical workloads	• Suitable for moderate workloads; shines in small-to-mid-sized setups	• Excellent performance in containerized workloads and hybrid cloud environments
	• High performance, but at higher costs	• Flexible and lightweight; performance depends on KVM efficiency	• Leading in hybrid cloud performance due to Kubernetes integration

This comparison in Table 7-1 highlights the strengths and suitability of each platform, enabling decision-makers to make informed choices based on their organization's unique requirements.

In summary, the decision to select between VMware, Proxmox, and RedHat necessitates a comprehensive understanding of an organization's specific requirements and strategic aspirations along with the budget. VMware continues to be a formidable option for enterprises with intricate, well-established infrastructures that demand high-quality, enterprise-level support. Conversely, for organizations aiming to minimize expenses, simplify management, and adopt more adaptable, cloud-native solutions, Proxmox and RedHat present attractive alternatives. Chief Experience Officers and IT executives must consider elements such as total cost of ownership, management complexity, and potential for future scalability to make well-informed choices that align with both present and future business goals.

7.3 Weighing Advantages and Disadvantages Across Vendors

When organizations assess various virtualization and cloud platforms, it is crucial to adopt a comprehensive perspective that considers the distinct advantages and disadvantages of each vendor. The selection of an appropriate platform transcends mere feature comparison; it necessitates an understanding of the strategic ramifications, long-term scalability, and potential obstacles associated with each option. Decision-makers and IT leaders must meticulously evaluate these elements to ensure the chosen platform meets both immediate requirements and long-term business goals.

Proxmox emerges as a notably cost-effective and customizable alternative, particularly suited for organizations that value flexibility and independence over the additional features offered by more traditional enterprise platforms. Its open-source framework grants businesses full control over their IT infrastructure, enabling them to customize configurations, integrate with existing systems, and avoid dependency on specific vendors. Furthermore, Proxmox's combination of KVM-based virtualization and Linux Containers (LXC) provides a cohesive solution for various workloads, thereby simplifying the management of both virtual machines and containerized applications.

For smaller enterprises or those with limited budgets, Proxmox presents an appealing choice as it eliminates the substantial licensing costs typically associated with proprietary platforms such as VMware. The capability to implement a comprehensive virtualization solution without a considerable financial outlay is particularly attractive for startups, educational institutions, and organizations undergoing digital transformation. Additionally, Proxmox features an intuitive, web-based management interface, which streamlines daily operations for IT administrators who may lack extensive experience in managing virtualization environments.

Proxmox, while a robust platform, has a notable limitation in its lack of enterprise-level support. Although it is supported by a vibrant community of users and open-source advocates, larger organizations may discover that this community-based assistance falls short of meeting the requirements for mission-critical applications or providing the necessary assurances for compliance and uptime. In environments where service-level agreements (SLAs) are crucial, the absence of formal support from a dedicated vendor can pose a significant challenge. Organizations that depend on round-the-clock support, especially those in highly regulated sectors like healthcare or finance, may find Proxmox's support offerings insufficient for their operational needs.

In contrast, RedHat offers a more established and enterprise-oriented solution, particularly suitable for businesses transitioning to hybrid cloud or multi-cloud frameworks. RedHat's OpenShift platform stands out in the realm of containerization and cloud-native operations, leveraging Kubernetes, which is recognized as the standard for container orchestration. RedHat is particularly effective in settings where organizations seek to modernize their application architectures, especially when implementing DevOps methodologies or adopting continuous integration/continuous delivery (CI/CD) practices. The platform's extensive integration capabilities with both private and public clouds facilitate seamless scaling across various environments, making it an ideal choice for companies with hybrid cloud aspirations.

RedHat's enterprise support ecosystem presents a considerable advantage for larger organizations. With certified Service Level Agreements (SLAs), dedicated support channels, and a global support team, RedHat ensures the stability and reliability essential for businesses managing critical workloads. For sectors such as banking, insurance, or government, where compliance, security, and reliability are paramount, RedHat's comprehensive support infrastructure offers the reassurance needed to confidently transition to a new platform.

On the other hand, a notable disadvantage of RedHat is the requirement for a more substantial investment, both financially and in terms of skill enhancement. Transitioning to RedHat, especially OpenShift, may necessitate retraining personnel, reassessing existing workflows, and investing in new infrastructure to leverage the platform's advanced features. Although the initial financial commitment may be considerable, the long-term benefits can be significant for organizations aiming to future-proof their operations and adopt cloud-native architectures. However, for smaller enterprises or those with constrained budgets, this upfront investment may pose a significant barrier.

VMware, a longstanding leader in virtualization, continues to be a feature-rich and stable option for enterprises heavily invested in traditional IT environments. VMware's extensive suite of tools for virtualization, networking, and storage management positions it as a preferred platform for companies requiring highly reliable and integrated infrastructure. vSphere, VMware's flagship offering, is recognized for its scalability, user-friendliness, and extensive third-party integrations, enabling organizations to manage complex workloads across various data centers and cloud environments. Furthermore, VMware's vSAN and NSX solutions deliver integrated storage and network virtualization, empowering companies to oversee their entire infrastructure from a unified platform.

The primary issue with VMware is its proprietary nature. Although VMware is proficient in managing traditional infrastructure, this model often results in vendor lock-in, complicating the transition to more open-source or cloud-native solutions. As businesses increasingly adopt multi-cloud strategies or aim to integrate with public cloud providers such as AWS, Azure, or Google Cloud, VMware's dependence on its own ecosystem can pose challenges. Moving away from VMware can be technically intricate and financially burdensome, necessitating significant migration efforts and meticulous planning to prevent disruptions to essential business functions.

Additionally, while VMware's licensing model is supported by its extensive feature set, it can be costly, especially for expanding businesses that require rapid scaling. The total cost of ownership (TCO) for VMware environments can rise quickly, including licensing fees, support contracts, and specialized hardware needs. For organizations with limited budgets, the high costs associated with VMware may restrict their capacity to invest in innovation or new technologies, forcing them to choose between maintaining outdated systems or embracing more contemporary solutions.

Ultimately, decision-makers must carefully consider these trade-offs when determining which platform best meets their organization's requirements. Proxmox provides cost-effectiveness and customization but lacks the formal support necessary for mission-critical environments. RedHat is strong in hybrid cloud integration and enterprise-level support but demands a considerable investment of time and resources. While VMware presents a robust and feature-rich platform, it can create financial pressure and vendor lock-in, thereby constraining future flexibility.

For organizations that value adaptability, cost efficiency, and the capacity for innovation, Proxmox or RedHat could be the most advantageous options. Conversely, companies that require stability, robust enterprise support, and a solution capable of handling intricate workloads at scale might discover that remaining with VMware provides the most immediate advantages, even if it entails higher expenses. By comprehending the strengths and weaknesses of each platform, CXOs can make well-informed choices that prepare their organizations for success in a progressively intricate IT environment.

7.4 Integration with Existing Infrastructure

When evaluating Proxmox as a viable alternative to their existing VMware environments, businesses are frequently attracted to its flexibility and open-source characteristics. Unlike certain proprietary solutions that may necessitate a complete migration, Proxmox can be implemented alongside existing VMware infrastructure. This presents a considerable advantage for organizations that depend on their VMware systems but wish to transition gradually to a more economical and open-source option without the upheaval of a large-scale migration. For example, companies can operate Proxmox in conjunction with VMware, leveraging KVM-based virtualization for specific workloads while still utilizing their VMware virtual machines for others, thus establishing a hybrid environment that combines the strengths of both platforms. However, it will double your management efforts at some point.

Consider a large organization with an extensive, multi-site infrastructure that utilizes VMware to manage numerous virtual machines (VMs). The prospect of completely overhauling this infrastructure in a single night is intimidating, not to mention the associated costs and risks. Instead, by implementing Proxmox alongside VMware, the organization can explore a gradual transition – starting with non-essential workloads or development environments on Proxmox. As familiarity with the new platform increases and cost efficiencies become evident, additional workloads can be migrated, facilitating a more strategic and phased approach to migration that minimizes disruptions to daily operations. Proxmox's integration features enable businesses to enjoy a smooth transition without the burden of a costly, high-risk replacement.

Conversely, RedHat provides a more enterprise-centric solution, particularly for organizations aiming for a hybrid cloud strategy that links their on-premises data centers with public cloud services such as AWS, Azure, or Google Cloud. RedHat's OpenShift, built on a robust Kubernetes foundation, supports seamless integration across both private and public cloud environments, making it an excellent choice for businesses that need flexibility and scalability across diverse settings. For instance, a company may have critical, highly sensitive workloads that must remain within their on-premises data centers for compliance purposes while simultaneously leveraging cloud-native services for development or data processing in public cloud environments.

In this context, RedHat's hybrid cloud model proves to be essential. By incorporating OpenShift into their legacy systems, organizations can expand their infrastructure into the cloud without the need for a complete system overhaul. This integration facilitates a smooth transition from legacy applications to cloud-native environments, enabling the operation of containers both on-premises and in the cloud. RedHat's hybrid architecture enhances workload portability, allowing applications to shift between on-prem and cloud environments as business requirements evolve, all without the necessity of extensive re-architecture. This adaptability is vital for organizations aiming to ensure their IT infrastructure can respond to changing operational demands.

Take, for instance, a financial services firm that utilizes a combination of legacy mainframes and VMware-based virtual machines and is increasingly interested in public cloud solutions for big data analytics. With RedHat's OpenShift, they can implement cloud-native applications while still maintaining support for the legacy systems that are crucial to their daily operations. This approach allows for a gradual modernization process, enabling the company to pilot cloud-native applications, scale them, and integrate them with existing systems without experiencing downtime or jeopardizing compliance. Additionally, RedHat's enterprise-grade support provides businesses undergoing these transformations with the necessary expertise and assistance to address any challenges that may arise during the transition.

Proxmox and RedHat each pose distinct integration challenges. Proxmox, being an open-source platform, provides significant customization and flexibility; however, organizations may need to invest in skilled IT staff capable of managing multiple virtualization platforms at once. This complexity can be especially overwhelming for smaller businesses that do not have the in-house expertise to run both VMware and Proxmox environments together. Furthermore, companies may need to consider investing in automation tools to streamline the management of a hybrid infrastructure, ensuring smooth workload transitions between Proxmox and VMware while reducing inefficiencies and potential risks.

RedHat faces significant challenges primarily due to the steep learning curve involved in managing Kubernetes-based environments, particularly for organizations that have historically relied on virtualization platforms such as VMware. Integrating OpenShift with existing legacy systems may require substantial upfront investments in both infrastructure upgrades and workforce training. Additionally, companies need to carefully manage the migration of applications to cloud-native architectures, ensuring that security, compliance, and performance standards are consistently met throughout the process. Ultimately, the successful integration of either Proxmox or RedHat with current infrastructure depends on effective strategic planning, employee development, and the implementation of automation and monitoring tools to facilitate seamless operations across various platforms. Proxmox's flexibility and openness present an appealing choice for those aiming for cost efficiency and gradual migration, whereas RedHat's robust hybrid cloud capabilities provide a strong solution for enterprises focused on modernization while ensuring compliance and stability.

In either scenario, the foremost priority should be to minimize downtime and prevent interruptions to essential operations. By implementing effective integration strategies, organizations can transition seamlessly to updated IT infrastructures, taking advantage of the capabilities offered by Proxmox or RedHat while still maximizing their current investments in legacy systems. This strategy not only guarantees operational continuity but also equips companies to adeptly manage the challenges of a changing cloud environment, thereby opening up new avenues for agility, innovation, and growth.

7.5 Security and Compliance Considerations

In today's hyper-connected digital landscape, the prevalence of data breaches and sophisticated cyber threats underscores the critical importance of security and compliance. For organizations across various sectors, particularly those in heavily regulated fields such as healthcare, finance, and government, selecting an IT platform that meets their security needs is essential for safeguarding sensitive information and adhering to industry regulations. As businesses contemplate moving from traditional platforms like VMware to open-source options such as Proxmox or enterprise-oriented solutions like RedHat, it is vital to assess how these platforms handle security and compliance, ensuring they can fulfill the stringent requirements of contemporary enterprises.

Proxmox's open-source architecture stands out as a significant advantage, enabling organizations to customize security measures to fit their unique requirements. This transparency allows businesses to understand the system's operations fully, facilitating tailored security configurations, rapid vulnerability patching, and the potential to develop proprietary security features as needed. For organizations with a robust internal IT security team capable of managing these customizations, Proxmox offers a flexible and cost-effective solution to address security challenges. For instance, it includes a wide range of built-in tools such as backups, snapshots, high availability, and a firewall.

However, the flexibility that comes with Proxmox also places a greater burden on organizations to ensure they are adhering to compliance standards. Unlike proprietary platforms that often come pre-configured with compliance frameworks in place, with Proxmox, organizations must take the lead in implementing and maintaining security measures that align with industry regulations. This could mean additional investment in security auditing tools, regular penetration testing, and in-house expertise to manage these requirements. For instance, businesses operating in the healthcare sector, which must adhere to strict compliance regulations like HIPAA, or in finance, with frameworks like PCI DSS, will need to ensure that Proxmox's open architecture can be configured to meet these compliance standards without exposing vulnerabilities.

The organization must ensure that patient data is encrypted both in transit and at rest, access controls are tightly enforced, and auditing mechanisms are in place to track who accesses sensitive information. While Proxmox allows for all of this, it requires manual configuration and monitoring, meaning the healthcare provider must have a dedicated security team capable of overseeing these elements. If the organization lacks the internal expertise to continuously monitor and update security settings, they may face challenges meeting regulatory compliance, potentially exposing themselves to penalties or legal risks in the event of a breach.

In contrast, RedHat takes a different approach, providing enterprise-grade security features out of the box, making it an attractive option for companies looking for a more turnkey solution. RedHat's platform, especially through its OpenShift and RedHat Enterprise Linux (RHEL) offerings, comes pre-integrated with a variety of security and compliance tools designed to meet the needs of highly regulated industries. For example, RedHat's OpenShift platform is built with Kubernetes-native security features that allow for granular role-based access control (RBAC), container isolation, and automated security patching, ensuring that the platform remains compliant with regulations without requiring extensive manual intervention.

RedHat also offers certified security frameworks aligned with regulatory requirements such as the Federal Information Security Management Act (FISMA), HIPAA, GDPR, and ISO/IEC 27001, giving businesses in regulated sectors the confidence that they are operating in a secure environment. This makes RedHat a low-risk option for organizations that need to adhere to specific compliance mandates but may lack the internal expertise or resources to build these capabilities from scratch. RedHat's certified security modules ensure that organizations can focus on their core business objectives rather than worrying about maintaining compliance with industry regulations. Also, if we talk about the healthcare sector, it has faced numerous hacking attempts, even with the implementation of security measures mandated by the Health Insurance Portability and Accountability Act (HIPAA) since 1996. Cybercriminals have developed sophisticated methods to exploit vulnerabilities, capitalizing on the significant value of Protected Health Information (PHI) and causing substantial disruptions within the healthcare system.

For example, a global financial institution considering a shift from VMware to RedHat would benefit from the latter's built-in encryption, secure access controls, and compliance frameworks specifically designed for financial transactions and data privacy regulations like PCI DSS. RedHat's platform would enable the financial institution to continue operating securely across multiple geographies, knowing that their data protection measures meet regulatory requirements. This would also reduce the overhead associated with manually configuring and auditing their infrastructure, allowing them to focus on driving innovation in their cloud-native financial applications.

Moreover, RedHat's hybrid cloud capabilities mean that businesses can extend security beyond their on-premises environments into public cloud platforms like AWS, Azure, and Google Cloud while maintaining a consistent security posture across all environments. The ability to leverage RedHat's OpenShift to manage security policies across both on-premises and cloud environments ensures that data remains protected no matter where it resides, helping businesses meet the increasing demands of data sovereignty laws and cross-border data transfer regulations.

One of the advantages RedHat has over Proxmox, particularly in the security and compliance arena, is its vendor support and access to a global network of certified security experts. For organizations that don't have the in-house security expertise, RedHat provides extensive support options to help manage security vulnerabilities, provide rapid updates, and conduct security audits, giving businesses the peace of mind that their platform remains compliant with industry standards. This is particularly important for large enterprises or government agencies, where security breaches or compliance failures can result in severe financial penalties or reputational damage.

On the flip side, RedHat's robust security offerings and enterprise-grade support come with a higher cost compared to Proxmox. For some organizations, especially small to mid-sized businesses, the upfront costs associated with RedHat may seem prohibitive. These organizations may opt for Proxmox due to its lower price point, relying on third-party security solutions or in-house talent to meet compliance standards. However, the total cost of ownership should be carefully weighed against the potential risks of non-compliance and security breaches, which can often far outweigh the savings associated with a more budget-friendly solution.

Here is a comparison table highlighting the Security and compliance considerations between Proxmox and RedHat based on the provided text.

Table 7-2. *Proxmox vs RedHat security and compliance*

Feature	Proxmox	RedHat
Security Customization	Open-source, customizable security configurations. Allows rapid vulnerability patching and development of proprietary security features.	Pre-configured enterprise-grade security with built-in tools like RBAC, container isolation, and automated security patching.
Compliance Flexibility	Requires organizations to configure and maintain compliance measures. Suited for those with strong internal IT security teams.	Comes with certified security frameworks (FISMA, HIPAA, GDPR, ISO/IEC 27001) out of the box, reducing the need for manual setup.

(continued)

Table 7-2. (*continued*)

Feature	Proxmox	RedHat
Vendor Support	Limited vendor support, reliance on community-based assistance. Organizations need in-house expertise for compliance.	Extensive global support network, certified security experts available to manage vulnerabilities and provide rapid updates.
Cost Consideration	Cost-effective, but requires additional investments in security auditing, penetration testing, and in-house expertise to maintain compliance.	Higher upfront cost, but reduces operational overhead for security management and compliance through built-in tools and support.
Best Use Case	Ideal for organizations with robust IT security teams capable of managing custom security measures or those needing a low-cost solution.	Suited for heavily regulated industries (e.g., finance, healthcare) where compliance and security are critical, and external support is required.
Hybrid Cloud Security	Requires manual configuration to extend security across on-premises and cloud environments.	Hybrid cloud capabilities with consistent security across on-premises and public cloud (AWS, Azure, Google Cloud).
Potential Risks	Greater risk of non-compliance if security is not properly managed. High reliance on internal teams for updates and security audits.	Lower risk of compliance failure due to vendor-backed security frameworks and support. Higher total cost of ownership but more reliable for critical workloads.

Table 7-2 summarizes the core differences between Proxmox and RedHat in terms of security and compliance, making it easier for organizations to make an informed decision based on their specific needs and resources. Ultimately, the decision between Proxmox and RedHat, when it comes to security and compliance, depends on the specific needs and capabilities of the organization. Proxmox offers greater customization and cost-efficiency but requires a hands-on approach to ensure that security and compliance are maintained at the highest level. RedHat, by contrast, provides a more comprehensive, out-of-the-box solution, making it an ideal choice for organizations that need to adhere to strict regulatory standards and prefer a managed security environment. Whether the choice is driven by the need for cost savings or compliance guarantees, understanding the security implications of each platform is critical for making an informed, strategic decision.

7.6 Performance and Scalability

As businesses grow and evolve, the performance and scalability of their IT infrastructure become increasingly crucial. Modern organizations face the challenge of ensuring that their systems can handle increased workloads, adapt to changing market demands, and support innovation – all without incurring exorbitant costs or creating inefficiencies. In this context, evaluating the performance and scalability of various virtualization platforms, including Proxmox, RedHat, and VMware, is essential for decision-makers who need to align their IT strategies with their growth ambitions. Each of these platforms offers distinct capabilities, and understanding how they perform in diverse environments can significantly influence an organization's ability to scale effectively while maintaining high performance.

Proxmox stands out for its flexibility and cost efficiency in scaling IT infrastructures. As an open-source platform, it allows organizations to build and expand their environments incrementally without the high licensing fees associated with proprietary solutions like VMware. This is particularly appealing to small- and medium-sized enterprises (SMEs) or startups that need to scale their infrastructure without making large upfront investments. With Proxmox, businesses can start small and gradually add resources – whether they are virtual machines (VMs), storage, or networking capabilities – as their needs grow. For instance, a growing e-commerce company might start with a modest Proxmox deployment, and as their website traffic and sales volume increase, they can expand their infrastructure to meet the new demand without overhauling their entire system.

However, while Proxmox offers cost-efficient scalability, its performance under high-demand enterprise workloads might not match that of more robust platforms like RedHat or VMware. For organizations with mission-critical applications or those operating in sectors that demand low-latency, high-availability systems – such as finance, healthcare, or telecommunications – Proxmox's KVM-based virtualization may face limitations in handling extremely large workloads or ensuring the high performance needed during peak operational periods. This is where RedHat becomes an attractive alternative.

RedHat, particularly with its OpenShift platform, is designed to excel in environments that require enterprise-grade performance and scalability. RedHat's architecture is built to seamlessly integrate hybrid cloud environments, allowing businesses to scale horizontally across public and private clouds with ease. The platform's Kubernetes-native orchestration, combined with its support for containerized applications, provides an unmatched ability to manage large-scale workloads efficiently. For businesses looking to adopt cloud-native strategies, RedHat offers the performance needed to run highly scalable applications while ensuring fast provisioning, autoscaling, and container management across multiple clouds.

For example, consider a global logistics company that needs to process vast amounts of data from thousands of deliveries daily, spanning different regions and time zones. With RedHat's hybrid cloud capabilities, this organization can distribute its workloads across both on-premises data centers and public clouds like AWS or Azure, ensuring that peak loads are handled efficiently without compromising performance. During the busy holiday season, when the volume of shipments surges, RedHat's autoscaling features ensure that the system can expand to accommodate the increased traffic without the risk of downtime or degraded performance.

This type of scalability is crucial for organizations undergoing digital transformation or those anticipating rapid growth. RedHat's performance optimizations for Kubernetes and containers also mean that applications can run with minimal resource overhead, allowing businesses to get more out of their infrastructure investments while maintaining agility.

In contrast, VMware, traditionally known for its reliability and feature-rich offerings, provides a solid foundation for enterprise workloads, but at a higher cost and with more rigid scalability options compared to Proxmox or RedHat. VMware has long been a trusted choice for large enterprises with extensive IT budgets, thanks to its well-established ecosystem of virtualization tools and enterprise-grade features. VMware's ability to handle complex workloads with granular control over virtualization resources makes it an excellent option for organizations with highly specialized requirements or those deeply embedded in VMware's ecosystem.

However, VMware's approach to scalability can be more costly and time-consuming. Scaling VMware environments often involves purchasing additional licenses and making significant investments in hardware and software upgrades. For organizations looking to scale rapidly while keeping costs under control, VMware's licensing structure can become a financial bottleneck. Additionally, VMware's reliance on a more proprietary architecture can limit flexibility, especially for organizations looking to integrate multi-cloud environments or take advantage of open-source technologies.

A good example of this contrast can be seen in a financial services company operating in a competitive global market. For this company, high availability, fast transaction processing, and real-time analytics are critical. While VMware may offer a reliable platform for such intensive workloads, the need to scale quickly to handle increased transaction volumes during market fluctuations might be hindered by licensing fees and the lack of cloud-native features. RedHat, with its container orchestration and hybrid cloud capabilities, would allow this financial institution to scale more dynamically across both on-premises and cloud environments, ensuring that they can meet market demands without sacrificing performance or agility.

Proxmox, on the other hand, is an attractive option for organizations that prioritize cost savings over performance at scale. For businesses with moderate workloads or those in less regulated industries, Proxmox provides the scalability needed without the steep learning curve or high costs associated with enterprise-grade solutions. However, as businesses grow and their demands become more complex, they may find themselves outgrowing Proxmox's capabilities and needing to consider more scalable platforms like RedHat or VMware.

In summary, each platform offers distinct advantages when it comes to performance and scalability, but the right choice depends on the organization's specific needs. Proxmox excels in cost efficiency and flexibility, making it ideal for businesses that require a low-cost, open-source solution to scale gradually. RedHat is the preferred option for enterprises looking to adopt cloud-native strategies with seamless hybrid cloud integration, offering superior performance and scalability for complex workloads. VMware, while more costly, remains a reliable choice for businesses deeply invested in its ecosystem, though it may lack the flexibility of the newer, more cloud-focused platforms. Ultimately, decision-makers need to assess their current infrastructure, workload requirements, and future growth trajectories to select the platform that will best support their long-term success.

7.7 Support and Community

The effectiveness of a platform's support ecosystem can significantly influence organizations, especially during critical situations such as troubleshooting, system failures, or intricate migrations. When evaluating platforms like VMware, Proxmox, and RedHat, the availability and quality of support can differ greatly, affecting both the routine management of infrastructure and the resolution of urgent issues.

VMware is recognized for its comprehensive and well-organized support framework. With years of experience in the enterprise sector, VMware offers an extensive support network that addresses everything from initial installation to troubleshooting and scaling. Their support services range from on-demand customer assistance to dedicated technical account managers, ensuring that businesses with complex requirements receive tailored support. For larger organizations, this level of service provides reassurance, knowing they can depend on VMware's global team of specialists during critical situations. However, this high-quality support is associated with increased costs, contributing to VMware's reputation for elevated overall operational expenses.

In contrast, Proxmox operates on a community-driven support model, which, while highly cost-effective, presents both opportunities and challenges. Proxmox's open-source nature means that its support largely stems from its active community of developers and users. For organizations that have in-house IT teams comfortable with open-source platforms, this can be a highly flexible and valuable resource. The community forums, documentation, and shared repositories of scripts and tools are robust, often solving issues without the need for vendor intervention. However, the absence of formal, 24/7 enterprise-level support can be a drawback for organizations that require immediate, guaranteed resolution for critical issues. This model might not be suitable for every organization, especially those without the internal technical expertise to troubleshoot or resolve issues independently.

RedHat provides a balanced solution that merges the advantages of robust, enterprise-level support with the open-source adaptability that many organizations desire. Its support framework is well-established, featuring both vendor-supported assistance and an active open-source community. This dual strategy instills confidence in businesses, ensuring they can access professional, enterprise-grade support when necessary while also tapping into the creativity and collaboration that the open-source community fosters. The subscription-based support model from RedHat guarantees customers receive timely updates, security patches, and performance improvements, making it an appealing choice for organizations pursuing hybrid cloud or multi-cloud strategies. The combination of community resources and vendor support positions RedHat favorably for businesses seeking the innovation and flexibility of open-source solutions without compromising on the stability and backing of a paid service.

For decision-makers, grasping the intricacies of these support ecosystems is essential. Organizations with large IT teams might find Proxmox's community-oriented support model adequate, benefiting from the cost efficiencies associated with an open-source approach. Conversely, companies operating in highly regulated sectors or those managing complex, mission-critical systems may prefer VMware's assured enterprise support or RedHat's unique blend of open-source innovation and professional backing.

Ultimately, the decision hinges on the unique requirements and risk management capabilities of the organization. Entities prioritizing cost-effectiveness and adaptability might opt for community-driven solutions such as Proxmox, accepting a trade-off in guaranteed, direct vendor assistance. Conversely, organizations in sectors where downtime can lead to substantial financial losses may deem it essential to invest in enterprise support systems like those offered by VMware or RedHat.

7.8 Use Cases and Industry Adoption

When assessing platforms such as Proxmox and RedHat, it is crucial to analyze their strengths in particular industry applications and understand why various organizations opt for these solutions over traditional alternatives like VMware. Each platform addresses distinct business needs, and their selection often depends on considerations like cost, scalability, security, and integration features. To gain a comprehensive understanding of Proxmox and RedHat's capabilities, examining real-world applications can illustrate how companies have successfully utilized these technologies.

Proxmox has emerged as a preferred choice for small to medium-sized businesses (SMBs) in search of an affordable and adaptable virtualization solution. Its open-source framework enables organizations with limited budgets to bypass significant licensing costs, a vital factor for SMBs aiming to enhance their IT expenditures without sacrificing performance. For instance, a small healthcare organization looking to virtualize its infrastructure may choose Proxmox due to its robust KVM-based virtualization, containerization through LXC, and user-friendly management via a web interface. These attributes facilitate easy deployment, maintenance, and scaling, even for businesses lacking extensive IT resources. By using Proxmox, this healthcare provider could consolidate servers, lower hardware expenses, and optimize its infrastructure all while avoiding the recurring costs associated with proprietary solutions like VMware. Additionally, Proxmox's capability to operate in hybrid environments alongside existing systems allows organizations to gradually shift their workloads, reducing the risk of significant disruptions. In the realm of education, where budget constraints are frequently encountered, Proxmox has gained significant traction. Educational institutions, including universities and colleges, typically operate numerous virtual machines (VMs) to support student laboratories, research initiatives, and administrative functions. The straightforward deployment process of Proxmox, along with its robust community backing, positions it as a compelling choice for these institutions, allowing them to quickly establish VMs for diverse applications without incurring substantial expenses. For instance, a technical university might implement Proxmox clusters to create distinct environments for various departments, enabling the IT team to efficiently manage multiple workloads without the need for expensive enterprise-level support. The adaptability of Proxmox encourages experimentation with both virtual machines and containerization, thereby nurturing a learning atmosphere that accommodates a broad spectrum of academic and research requirements.

Conversely, RedHat excels in large enterprises and sectors that demand comprehensive hybrid cloud infrastructures and enterprise-grade support. Industries such as telecommunications, financial services, and government entities often face more rigorous demands regarding security, compliance, and scalability. RedHat, with its enterprise-level offerings like OpenShift, stands out in the field of Kubernetes-based container orchestration, empowering organizations to develop, deploy, and scale applications across hybrid cloud settings. For example, a global telecommunications provider may choose RedHat to enhance its infrastructure and optimize service delivery. By utilizing OpenShift, the provider could establish a cloud-native framework capable of managing vast volumes of data and applications distributed across multiple locations. The seamless integration with both on-premises data centers and public cloud platforms such as AWS and Azure offers the necessary agility for innovation while ensuring compliance and data protection.

The financial services industry, characterized by intricate regulatory frameworks and a strong emphasis on security, greatly benefits from RedHat's comprehensive solutions. For instance, a global banking institution may implement RedHat's hybrid cloud technologies to enhance operational efficiency across its international branches. By incorporating RedHat OpenShift into its systems, the bank can securely manage sensitive customer information within a compliant framework while utilizing cloud scalability for the development and rollout of new products. The robust support network provided by RedHat, along with its commitment to open-source advancements, allows the bank to remain agile in its operations while adhering to the rigorous standards set by regulatory bodies.

In the manufacturing sector, where the demand for real-time data processing and machine learning capabilities is on the rise, RedHat's offerings facilitate the transition to edge computing. Manufacturers require platforms that can seamlessly integrate with existing legacy systems while accommodating cutting-edge technologies such as AI and IoT. RedHat's capacity to scale from central data centers to edge environments, coupled with its focus on security and automation, positions it as an essential resource for modernizing manufacturing processes.

In conclusion, both Proxmox and RedHat serve distinct industry requirements and applications, establishing themselves as formidable alternatives to conventional solutions like VMware. Proxmox is particularly advantageous for small to medium-sized businesses and educational institutions seeking affordable virtualization options that offer flexibility and scalability. Conversely, RedHat is tailored for large enterprises operating in heavily regulated sectors, where hybrid cloud frameworks, extensive support, and high-performance computing are vital for achieving success. As organizations continue to adapt to evolving technological landscapes, these platforms present effective alternatives that strike a balance between innovation, cost-effectiveness, and enterprise-level reliability.

7.9 Open-Source vs. Proprietary Solutions

The discussion surrounding open-source versus proprietary solutions extends beyond mere cost considerations; it encompasses aspects such as control, adaptability, innovation, and risk management. Organizations have historically faced challenges in selecting between these two options, and as the IT environment grows increasingly intricate, the significance of this choice intensifies. Open-source platforms like Proxmox and RedHat offer avenues for customization, innovation, and cost efficiency. Conversely, proprietary leaders such as VMware deliver enterprise-grade support, refined user experiences, and a tightly regulated environment that reduces risk, albeit at a premium price. Fundamentally, open-source solutions such as Proxmox and RedHat grant businesses the liberty to innovate according to their own timelines. By providing access to the source code, these platforms enable organizations to modify their IT infrastructure to align with specific needs. For instance, a technology firm looking to develop a highly tailored application environment might choose Proxmox or RedHat, as these platforms facilitate extensive customization, allowing developers to adjust configurations, create unique solutions, and seamlessly integrate advanced tools. This level of flexibility is particularly advantageous for companies with proficient technical teams eager to maximize their IT capabilities.

Proxmox serves as a prime example of a robust solution, offering KVM-based virtualization alongside LXC container support, making it an ideal choice for organizations seeking lightweight and adaptable options without the constraints of vendor lock-in. Companies that emphasize agility and cost-effectiveness may find Proxmox's open-source framework particularly appealing, as it facilitates the creation of hybrid environments where virtual machines and containers can coexist seamlessly. Additionally, Proxmox's community-driven development approach fosters rapid innovation, with enhancements and new features often reflecting the community's requirements rather than corporate interests. This aspect can be particularly beneficial for small- to medium-sized businesses, startups, or research institutions that operate on limited budgets yet require flexibility and advanced solutions.

In a similar vein, RedHat, while more focused on enterprise needs, also embraces the open-source philosophy within the hybrid cloud landscape. RedHat's OpenShift, which is built on Kubernetes, offers enterprises an open-source platform that excels in hybrid and multi-cloud environments. By choosing RedHat, large organizations can utilize Kubernetes for container orchestration, enabling them to develop scalable, cloud-native applications that can transition effortlessly between public and private clouds. This ability to scale, paired with an open-source foundation, empowers companies in industries such as finance and telecommunications to remain at the cutting edge of technology without being constrained by the limitations of proprietary systems.

The freedom associated with open-source solutions is accompanied by significant responsibility. While these solutions are often groundbreaking, they demand a greater level of technical knowledge. Organizations may face a steep learning curve and might need to invest in training their IT personnel to effectively manage and utilize these systems. The very customization and flexibility that make open-source appealing can pose challenges if the necessary technical skills are not available. For instance, a healthcare organization implementing Proxmox may encounter difficulties in optimally configuring the platform to meet its specific requirements without in-house Linux expertise. Additionally, the lack of a strong vendor support network means that companies must depend largely on community resources for troubleshooting and updates, which, although valuable, may not provide the same level of responsiveness or assurances that a proprietary vendor like VMware can deliver.

Conversely, proprietary solutions such as VMware are adept at offering well-structured, reliable environments, instilling confidence in businesses that their infrastructure is fully backed by a vendor with extensive experience. VMware's vSphere and other virtualization products are tailored for enterprise clients, featuring ready-to-use capabilities that require minimal customization. This makes VMware particularly suitable for sectors like healthcare, government, and manufacturing, where compliance, security, and system uptime are critical. In these industries, the priority often lies in ensuring that the platform operates smoothly from the outset rather than in the ability to modify it extensively.

VMware's proprietary model presents several trade-offs. A primary concern is the elevated costs linked to licensing and support agreements. When compared to alternatives like Proxmox or RedHat, VMware's pricing can be a barrier, particularly for small- to medium-sized businesses or those with limited IT budgets. Moreover, proprietary solutions often lead to vendor lock-in, making organizations increasingly reliant on a single provider's suite of tools, which restricts their ability to embrace newer, innovative technologies outside that ecosystem. This dependency can pose challenges for companies aiming to diversify their infrastructure or transition to multi-cloud strategies.

Additionally, while VMware offers robust enterprise support, it may lack the adaptability that open-source platforms provide. Organizations looking to experiment, innovate, or incorporate advanced open-source technologies may find VMware's proprietary limitations to be a hindrance. For instance, a financial services company might wish to implement open-source solutions like Prometheus for monitoring or Ansible for automation, but achieving seamless integration within a VMware-focused infrastructure could be challenging without sacrificing cost or performance.

When evaluating the two methodologies, decision-makers need to consider the total cost of ownership (TCO), which encompasses licensing, support, and maintenance expenses, in relation to the long-term strategic benefits of flexibility and innovation. In certain instances, the initial cost savings associated with open-source options such as Proxmox or RedHat can be redirected toward innovation, enabling organizations to concentrate on enhancing new capabilities instead of managing licenses. Conversely, the assurance provided by VMware's enterprise-level support and reliability may justify the expense, particularly in sectors where downtime or compliance issues are intolerable.

Ultimately, the choice between open-source and proprietary solutions hinges on the organization's long-term goals, technical expertise, and financial limitations. Companies seeking autonomy, cost efficiency, and the potential for innovation are likely to favor Proxmox or RedHat. In contrast, those emphasizing stability, ease of management, and robust enterprise support may find VMware to be a more appropriate option. Regardless of the selected route, it is essential to comprehend the trade-offs between customization, cost, and control to make a well-informed decision that aligns with the organization's requirements and future growth aspirations.

7.10 Automation and Orchestration

Proxmox, recognized for its lightweight and adaptable architecture, integrates seamlessly with widely used automation tools like Ansible. This makes it a compelling option for organizations that are already utilizing DevOps pipelines or looking to embrace Infrastructure-as-Code (IaC) methodologies. Ansible's declarative language empowers teams to articulate infrastructure and deployment configurations as code, which can be version-controlled, reused, and automated across multiple environments. By automating processes such as provisioning, configuration management, and application deployment, Proxmox and Ansible help IT teams reduce manual workloads and standardize operations. This approach is particularly beneficial for small- and medium-sized businesses (SMBs) that may lack dedicated infrastructure management teams but still aim to harness automation for enhanced efficiency.

In this scenario, Proxmox stands out by enabling users to develop adaptable and tailored automation workflows without being confined to a proprietary framework. Its compatibility with API-driven orchestration tools empowers organizations to construct and automate processes that cater to their specific requirements. For instance, a media organization could utilize Proxmox to oversee virtual machines (VMs) that facilitate their content production workflows, with Ansible handling the scaling and configuration of these VMs in response to fluctuating workloads. This degree of customization enables businesses to swiftly adjust to evolving demands while maintaining cost efficiency.

Conversely, RedHat has established itself as a frontrunner in delivering powerful automation solutions, especially for enterprises functioning within intricate, hybrid cloud settings. RedHat's OpenShift, which is based on Kubernetes, provides a robust platform for automating not only conventional infrastructure operations but also the management of containerized applications at scale. OpenShift features integrated automation functionalities that streamline the deployment, scaling, and oversight of containerized workloads across both on-premises and cloud environments. This makes RedHat particularly appealing to large enterprises or sectors that require the management of highly distributed and scalable applications, such as finance, healthcare, and telecommunications.

A key benefit of RedHat OpenShift lies in its robust integration with Kubernetes orchestration, which facilitates automated load balancing, self-healing, and rollbacks for containerized applications. OpenShift enhances Kubernetes' inherent automation features by incorporating enterprise-level functionalities such as policy-driven management, improved security, and support for intricate CI/CD pipelines (continuous integration/continuous deployment). This integration simplifies the management of large-scale application deployments for DevOps teams, ensuring consistent updates across hybrid or multi-cloud environments. Organizations aiming to automate the lifecycle of containerized applications – from development to production – will discover that OpenShift's automation features foster increased agility, quicker time to market, and enhanced reliability.

VMware, a well-established leader in virtualization, offers a comprehensive array of automation and orchestration tools. VMware vRealize Automation (vRA) and vSphere empower organizations to automate the deployment and management of virtual machines, applications, and infrastructure services. With VMware, companies can utilize features such as policy-based automation, self-service portals, and integrated monitoring to oversee both virtualized environments and hybrid cloud infrastructures. VMware's solutions are highly refined, providing deep integrations with other VMware products and a level of enterprise support that is difficult to find in the open-source domain. However, while VMware's automation tools are powerful, they often come with higher licensing fees, which may not be ideal for organizations seeking more budget-friendly options.

VMware offers numerous advantages; however, Proxmox and RedHat present unique benefits stemming from their open-source nature. Organizations that value flexibility and customization can leverage open-source automation for enhanced control over their infrastructure management. Proxmox enables users to create tailored automation workflows that align precisely with their requirements, free from proprietary limitations. The community-driven approach to developing automation tools allows Proxmox to swiftly respond to emerging industry trends, providing users access to a wealth of community-generated scripts and templates that can be adapted for specific applications.

On the other hand, RedHat merges the flexibility of open-source automation with the dependability of enterprise-grade support, giving it a competitive edge in sectors where scalability and compliance are paramount. RedHat's Ansible Tower, an advanced iteration of the open-source Ansible, equips enterprises with centralized management, role-based access, and comprehensive reporting features, facilitating the automation of even the most intricate IT environments. For organizations in regulated industries like finance or healthcare, these additional capabilities are crucial for ensuring that automated processes are both efficient and compliant with industry regulations.

As IT infrastructure continues to advance, the significance of automation and orchestration in enhancing operational efficiency and agility is set to increase. Organizations that prioritize automation now will find themselves in a stronger position to expand their operations, adapt to market changes, and surpass their competitors. Proxmox, RedHat, and VMware each provide unique approaches to automation, tailored to various business requirements and technical settings. Proxmox stands out for its affordability and high degree of customization, making it attractive for businesses seeking a versatile solution that integrates seamlessly with existing DevOps processes. RedHat excels in hybrid cloud capabilities and enterprise-level automation, making it suitable for organizations that need to manage complex applications across diverse environments. Meanwhile, VMware, known for its comprehensive tools and strong support network, remains a preferred option for enterprises looking for a feature-rich, ready-to-use automation platform.

The choice of a suitable automation and orchestration platform will ultimately depend on the unique goals and operational requirements of the organization. Whether opting for the flexible open-source options provided by Proxmox and RedHat or the dependable enterprise solutions from VMware, companies that prioritize automation are likely to experience improved efficiency, scalability, and innovation within their IT operations.

As the digital environment evolves rapidly, it is imperative for organizations to stay ahead to maintain their competitive advantage. The swift progress in cloud computing, virtualization, and automation is transforming business operations, making it crucial for leaders to remain aware of emerging trends. Proxmox and RedHat are strategically positioned to adapt to these innovations, delivering advanced solutions that meet the future demands of IT infrastructure. This section delves into the key trends and innovations shaping the future of these platforms and underscores the importance of understanding them for organizations considering long-term investments.

One of the most notable transformations in IT infrastructure is the emergence of edge computing. As IoT devices become more widespread and the need for real-time data processing intensifies, edge computing has established itself as an essential element of contemporary infrastructure. By processing data closer to its source, edge computing minimizes latency and facilitates quicker decision-making. Proxmox, known for its lightweight design and adaptability, is particularly advantageous for edge implementations. Its capability to operate on smaller, less resource-demanding hardware makes it perfect for organizations aiming to set up micro data centers or edge nodes in remote areas. Additionally, Proxmox's open-source framework allows for extensive customization, enabling companies to optimize their edge environments for specific workloads without the complexities associated with traditional virtualization solutions. Conversely, RedHat is making notable advancements in edge computing with its OpenShift platform. The rise of containerization has positioned RedHat's Kubernetes-based architecture as an excellent solution for managing distributed workloads across both cloud and edge settings. RedHat's emphasis on hybrid cloud solutions allows businesses to effortlessly extend their cloud operations to the edge, ensuring a cohesive and scalable infrastructure. As organizations seek to leverage the advantages of edge computing, RedHat's enterprise-level support and strong security features position it as a preferred option for sectors such as telecommunications, manufacturing, and healthcare, where real-time data processing is crucial.

A significant trend influencing the future of IT infrastructure is the growing importance of artificial intelligence (AI) and machine learning (ML) in the management and optimization of systems. AI-enhanced infrastructure management enables organizations to automate intricate decision-making processes, improve resource allocation, and foresee potential system failures before they arise. Proxmox, as an open-source platform, offers the adaptability to incorporate AI and ML algorithms into its infrastructure. Organizations can utilize AI-driven automation tools to improve system monitoring, anticipate workload requirements, and dynamically allocate resources, all while minimizing operational expenses.

Similarly, RedHat is making substantial investments in AI-driven automation through its Ansible Automation Platform and OpenShift. By focusing on the integration of AI within its cloud-native infrastructure, RedHat empowers businesses to develop intelligent, self-managing environments. For instance, AI-enabled orchestration tools in OpenShift can automatically adjust the scaling of containerized applications in response to real-time workload demands, ensuring both optimal performance and cost-effectiveness. Furthermore, RedHat's application of AI and ML extends to security, featuring AI-based threat detection systems capable of identifying and addressing potential vulnerabilities before they escalate into critical problems. As organizations increasingly embrace AI to foster innovation, RedHat's proactive strategy positions it as a frontrunner in intelligent infrastructure management.

One of the most significant developments in IT infrastructure is the transition toward hyperconverged infrastructure (HCI) and software-defined everything (SDx). HCI streamlines data center management by combining computing, storage, and networking into a single platform that is controlled through software rather than hardware. Proxmox is an excellent choice for organizations aiming to implement HCI, providing a budget-friendly, software-defined virtualization solution. Its integration of KVM and LXC containers enables companies to consolidate their resources and manage them via a cohesive interface, thereby minimizing complexity and operational costs.

RedHat is also making notable advancements in the HCI domain with its OpenShift Virtualization platform. By merging virtual machines and containers into one unified system, OpenShift allows organizations to oversee both legacy and cloud-native applications through a consistent, software-defined framework. This integration facilitates a seamless transition from traditional virtualized setups to containerized environments, enabling businesses to modernize their infrastructure without interrupting ongoing operations. As organizations strive to simplify their data center processes and lessen their reliance on hardware, the adoption of HCI and SDx is expected to rise, with both Proxmox and RedHat leading the way in this evolution.

The increasing focus on multi-cloud and hybrid cloud environments is a significant trend propelling innovation in IT infrastructure. Organizations are progressively implementing multi-cloud strategies to mitigate vendor lock-in, enhance cost efficiency, and guarantee high availability. RedHat, with its robust emphasis on hybrid cloud solutions, empowers businesses to manage workloads effortlessly across public, private, and edge clouds. OpenShift serves as a cohesive platform for orchestrating containerized applications across various cloud providers, enabling organizations to capitalize on the unique advantages of each provider while retaining control over their data and applications.

Proxmox, primarily oriented toward on-premises and private cloud setups, also facilitates multi-cloud integration by managing both virtual machines and containers through a single interface. Its compatibility with a range of cloud storage providers, along with its open-source framework, allows organizations to create tailored multi-cloud environments that meet their specific requirements. As businesses continue to broaden their cloud strategies, the capability to integrate and manage workloads across diverse cloud platforms will become increasingly vital.

Organizations continue to prioritize security as they embrace new technologies and infrastructure models. Proxmox and RedHat are both making strides in enhancing security and compliance, particularly through automated security management. Proxmox's open-source platform enables companies to tailor their security measures to align with specific industry standards, while RedHat provides enterprise-level security features that include built-in compliance tools, facilitating easier adherence to regulations. The incorporation of AI-driven security solutions will further empower organizations to identify and address threats in real-time, ensuring the integrity of their infrastructure in an increasingly intricate digital environment.

Moreover, sustainability is emerging as a crucial factor for businesses assessing their IT infrastructure investments. Green IT initiatives focused on minimizing the environmental footprint of data centers are fostering advancements in energy-efficient computing. Both Proxmox and RedHat are dedicated to promoting sustainable IT practices through their open-source and hybrid cloud solutions, which enable organizations to optimize resource utilization and decrease energy consumption. As companies grow more environmentally aware, investing in infrastructure that champions sustainability will not only lower operational expenses but also align with their corporate social responsibility objectives.

The future of IT infrastructure is being significantly influenced by a blend of emerging trends and innovations that are redefining business operations. Proxmox and RedHat are at the forefront of this evolution, providing adaptable, scalable, and secure solutions tailored to the requirements of contemporary organizations. With the increasing prominence of edge computing, AI-enhanced infrastructure management, and the shift toward hyperconverged infrastructure and multi-cloud environments, these platforms are continuously evolving to address the challenges of a fast-paced digital landscape. For organizations considering long-term investments, grasping these trends is essential for choosing the appropriate platform that aligns with their growth and innovation objectives. Whether leveraging the flexibility and cost-effectiveness of Proxmox or the robust capabilities of RedHat, businesses have access to a variety of powerful resources to ensure their IT infrastructure remains resilient and future-ready.

7.11 Examples for References

Here, I have suggested some possible pathways and some hypothetical examples. However, they are not direct reflections of actual industry cases. Instead, think of them as inspiration to tailor your own approach, ensuring alignment with your unique business needs and IT landscape as you embark on your modernization journey. This section explores comprehensive examples that illustrate how various organizations can possibly and effectively move from conventional virtualization platforms to alternatives such as Proxmox and RedHat. These narratives offer important insights into the real-world implementation of these technologies, emphasizing both the benefits realized and the challenges faced during the transition. For decision-makers considering a similar path, these examples present valuable lessons, best practices, and strategic insights that can guide their decision-making process.

7.11.1 Example 1: XTech – A Global Manufacturing Leader Adopts Proxmox for Cost Efficiency and Flexibility

XTech, a global manufacturing firm, had depended on VMware for over 10 years to oversee its virtualization framework. As the company grew internationally, its infrastructure expanded correspondingly, resulting in increased operational expenses, heightened complexity, and frequent maintenance interruptions. Although VMware had initially met their needs, XTech's IT leadership acknowledged that their current infrastructure was becoming neither cost-effective nor adaptable enough to support the requirements of a rapidly expanding business. Notably, scaling their VMware environment often entailed substantial licensing costs and considerable hardware investments.

In response, XTech began investigating open-source solutions and ultimately opted to migrate a significant portion of its operations to the Proxmox Virtual Environment. The key motivations for this shift included cost savings, enhanced flexibility, and the ability to customize. The open-source model of Proxmox removed the burden of costly licensing fees, enabling XTech to allocate more resources toward innovation and digital transformation projects. Additionally, Proxmox's compatibility with KVM-based virtualization facilitated smooth integration with the existing infrastructure, thereby reducing downtime during the transition.

The transition presented several challenges, especially regarding the need to enhance the IT team's skills for managing an open-source platform. However, the advantages soon became evident. The company successfully implemented Proxmox clusters throughout its global operations, resulting in a 20% decrease in operational costs within the first year. Furthermore, Proxmox's efficient architecture facilitated more adaptable scaling, enabling XTech to grow its data centers in line with business expansion without the significant expenses typically associated with proprietary solutions.

Looking back, XTech's leadership considers the shift to Proxmox a crucial milestone in the company's digital transformation journey. The key takeaway from XTech's experience is unmistakable: open-source solutions like Proxmox can provide a cost-effective and scalable alternative to proprietary platforms, as long as organizations commit to developing the necessary skills and carefully plan the transition to minimize risks.

7.11.2 Example 2: Telecom Titan Transforms Hybrid Cloud Operations with RedHat OpenShift

A prominent telecommunications firm, TeleTech, reached a pivotal moment in its IT modernization efforts. With a vast and distributed infrastructure catering to millions of customers, the organization faced challenges in managing both on-premises systems and a swiftly growing cloud presence. Their current VMware setup had become inflexible and expensive, making it difficult to scale effectively across public and private clouds. Furthermore, as containerized applications and cloud-native development gained significance, the necessity for a platform capable of accommodating these changing workloads became clear.

After a thorough assessment of various alternatives, TeleTech opted to transition its infrastructure to RedHat OpenShift, a leading solution for hybrid cloud and container orchestration. This choice was influenced by OpenShift's strong integration with Kubernetes, facilitating the efficient management of containerized applications in both on-premises and cloud settings. The company also appreciated RedHat's enterprise-level support and the comprehensive security features that aligned with their rigorous compliance standards in the telecommunications industry.

The transition to OpenShift presented several challenges for TeleTech. The company needed to make substantial investments in upskilling, ensuring that its IT and DevOps teams were well-equipped to navigate the new environment. Furthermore, the migration of essential applications from VMware to a containerized setup demanded careful planning and a phased approach to prevent any service disruptions. Nevertheless, TeleTech successfully completed this transition in just over a year.

The advantages gained were significant. By utilizing OpenShift, TeleTech achieved a 40% reduction in time-to-market for new services, driven by enhanced efficiency in its DevOps pipelines and the automation of routine tasks such as infrastructure provisioning. Additionally, OpenShift's Kubernetes-native framework enabled the company to scale its operations flexibly, responding to varying customer demands without incurring substantial costs.

The primary lesson from TeleTech's experience highlights the effectiveness of hybrid cloud environments in enhancing operational agility. For organizations seeking a versatile, cloud-native platform that robustly supports enterprise workloads, RedHat OpenShift presents a strong solution that harmonizes innovation with dependability.

7.11.3 Example 3: Financial Services Giant Leverages Multi-cloud Flexibility with Proxmox

A global financial services firm, FinCo, encountered an increasingly intricate IT environment as the need for scalability, security, and high availability intensified. Although their VMware setup had been effective for many years, they began experiencing challenges related to cost control and the constraints of vendor lock-in, especially as they aimed to integrate with various cloud providers such as AWS, Azure, and Google Cloud.

To address these issues, FinCo adopted a multi-cloud strategy, incorporating Proxmox as a fundamental element of its on-premises infrastructure. The open-source nature of Proxmox provided the company with the ability to manage both virtual machines and containers within a unified environment, offering enhanced flexibility compared to their previous VMware solution. Additionally, Proxmox's compatibility with multiple cloud platforms enabled FinCo to seamlessly connect its on-premises systems to different cloud providers, thereby improving disaster recovery options and optimizing resource distribution across clouds.

The transition to Proxmox was executed in phases, beginning with the deployment of Proxmox clusters in non-production settings to verify compatibility and stability. Following the successful completion of these pilot projects, the migration progressed to mission-critical systems. This transition allowed FinCo to lessen its dependence on VMware's proprietary ecosystem, granting the company greater flexibility in managing its infrastructure.

FinCo achieved a 30% reduction in total cost of ownership (TCO) and enhanced its operational agility, enabling the company to seamlessly transfer workloads among cloud providers as necessary. This adaptability was particularly beneficial during times of significant market fluctuations, when there was an unexpected spike in infrastructure demand. FinCo's capacity to rapidly scale and reallocate resources ensured consistent service availability and performance, even in challenging conditions.

The key takeaway from FinCo's experience is that Proxmox's multi-cloud flexibility and cost-effectiveness equip financial services organizations with essential tools to modernize their infrastructure while reducing the risks linked to vendor lock-in.

These examples can illustrate the diverse strategies organizations can adopt when moving to alternative virtualization platforms such as Proxmox and RedHat. Each organization encountered unique challenges – ranging from cost management and scalability to the necessity for hybrid cloud integration – but ultimately discovered solutions that not only met their immediate requirements but also established a foundation for future expansion. For decision-makers contemplating similar transitions, these narratives offer practical insights, emphasizing the significance of thorough planning, skill enhancement, and a clear understanding of the distinct advantages each platform offers.

7.12 Summary

As organizations adapt to the changing dynamics of IT infrastructure, dependence on a single vendor, such as VMware, presents both operational and financial hurdles. This chapter examined how alternative platforms like Proxmox and RedHat provide attractive options for companies in search of flexibility, scalability, and cost-effectiveness. Proxmox, with its open-source model and affordability, is particularly suited for small to medium-sized enterprises, whereas RedHat's robust security features and hybrid cloud capabilities cater to larger organizations. The analysis in this chapter emphasizes that the optimal choice is contingent upon an organization's size, regulatory requirements, and strategic goals.

For executives and decision-makers, the takeaway is evident: adopting a multi-vendor IT approach can improve agility, lower expenses, and reduce the risks linked to vendor dependency. While VMware is proficient in offering a comprehensive enterprise solution, Proxmox and RedHat empower businesses to diversify, innovate, and prepare for future challenges. Security and compliance are paramount, with RedHat delivering strong, enterprise-level security and Proxmox providing the adaptability for tailored security measures.

In summary, making well-informed decisions regarding IT infrastructure is crucial for sustained success. By considering alternatives like Proxmox and RedHat, organizations can enhance resilience, optimize performance, and ensure their infrastructure is ready to meet future demands.

CHAPTER 8

Author Recommendations and Insights

As we delve into the realm of virtualization and cloud infrastructure, we find ourselves at a significant juncture in technological advancement. Organizations today face crucial decisions about whether to stay with VMware, transition to Proxmox, explore Azure Stack HCI, or adopt RedHat, or leverage the innovative capabilities of Nutanix. Each option presents distinct advantages, challenges, and potential for growth. This chapter examines these essential factors, providing guidance for navigating the complex virtualization environment. The pressing question remains: Which choice will foster unmatched agility, cost efficiency, and sustained innovation? Let's explore the future of cloud-driven enterprises.

In this segment, we shift from analysis to a practical approach. Drawing on our extensive experience with various platforms and numerous real-world scenarios, we will share our recommendations and insights that go beyond technology to address the broader implications for organizations across different industries. We will begin by delving into the often-overlooked features of hyper-converged infrastructure (HCI) and Microsoft's Hyper-V, both of which are emerging as viable alternatives in this swiftly changing environment. Additionally, Microsoft's upcoming

© Sumit Bhatia and Chetan Gabhane 2025
S. Bhatia and C. Gabhane, *Navigating VMware Turmoil in the Broadcom Era,*
https://doi.org/10.1007/979-8-8688-1264-4_8

innovations with Windows Server 2025, a significant shift in our approach to virtualization, security, and multi-cloud strategies, which we believe will play a crucial role in shaping the future of IT ecosystems.

Consider this chapter as a valuable resource designed to help you traverse the complex array of choices at your disposal. While previous chapters have thoroughly examined aspects such as costs, management, security, and suitability, this section will concentrate on providing you with practical insight addressing the question of "what's next?" By analyzing various scenarios and use cases, you will learn how different platforms are utilized across diverse sectors, including finance, healthcare, retail, and manufacturing. Although there is no universal solution, there are strategic pathways that can inform your decisions, tailored to your organization's specific needs, challenges, and growth plans.

Imagine a medium-sized retail company ready to embrace digital transformation. They are grappling with rising costs and complexities tied to VMware and are unsure whether to switch to open-source options like Proxmox or to a more enterprise-focused solution such as RedHat. We will explore key lessons from different industries, evaluate best practices for managing multi-cloud setups, and offer tailored insights that cater to businesses of all sizes, from budding startups to established corporations. On the other hand, large healthcare organizations that needs to comply with strict regulatory requirements while handling sensitive data. Here, security, compliance, and reliability are crucial. These organizations must assess their current circumstances and plan for the future, particularly as IT infrastructure evolves with trends such as edge computing, containerization, and hybrid-cloud models.

This chapter takes a broader view than just the technology; it highlights the importance of incorporating these tools into long-term business strategies, using virtualization as a driver for innovation instead of just a necessity for operations. We will explore valuable lessons from companies that have effectively embraced alternative solutions, examining their hurdles and achievements while uncovering key insights that could benefit your organization. Whether you're a technical decision-maker looking to boost ROI or an IT leader striving to achieve a balance between performance and scalability, these practical examples will offer crucial perspectives.

8.1 A Strategic Perspective for Private Cloud with Hyper-V HCI

If you're building or optimizing a private cloud, HCI with Hyper-V should be at the top of your list. Its unified approach simplifies infrastructure management, its scalability supports growth, and its security and resilience align perfectly with private cloud demands. This combination provides the flexibility and reliability businesses need to stay competitive in a cloud-driven world. It does not require Azure connectivity, but it can integrate with Azure hybrid services for cloud-based monitoring, backup, recovery, and management.

8.1.1 The Strategic Advantage of HCI and Hyper-V

When utilizing Hyper-V alongside HCI to create your private cloud solution, it is essential to take into account the architecture described below. Hyper-V combined with HCI is an excellent option for organizations that emphasize strong on-premises virtualization while minimizing dependence on hybrid cloud capabilities. It's worth mentioning that although Hyper-V operates autonomously, it can effortlessly connect with Azure to access enhanced features like backup, monitoring, and site recovery. Please refer Figure 8-1 which illustrates Windows Admin Center setup with Storage Spaces Direct. Here, a storage pool is connected to servers via a switchless network using two 25G NICs. The servers run Hyper-V, Software Defined Storage (SDS), and Software Defined Network (SDN) on Microsoft Windows Server HCI OS.

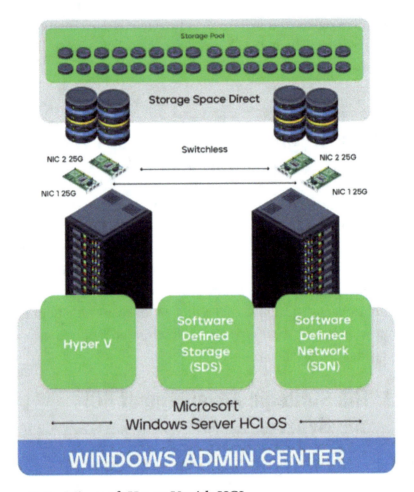

Figure 8-1. *Microsoft Hyper-V with HCI*

HCI integrates computing, storage, and networking into one cohesive solution, simplifying management and enhancing scalability. Microsoft's virtualization platform, Hyper-V, enhances this by providing strong virtual machine management and smooth integration with HCI capabilities such as Storage Spaces Direct (S2D). This combination creates a dependable infrastructure for private clouds, perfect for organizations that want to maintain control over their IT environments while being flexible enough to accommodate hybrid deployments.

8.1.2 Azure Stack HCI: How It Differs from Hyper-V?

In Chapter 3, we have explored the details about Azure Stack HCI (hyper-converged infrastructure). Azure Stack HCI and Hyper-V, both from Microsoft, play complementary but distinct roles in today's IT landscape. Hyper-V serves as a powerful virtualization platform, while Azure Stack HCI takes this foundation further by enabling a seamless hybrid cloud experience. If your organization aims to modernize its infrastructure, integrate effortlessly with Azure services, or support hybrid workloads and edge computing, Azure Stack HCI provides a future-ready solution. It bridges the gap between on-premises and cloud environments, delivering scalability, flexibility, and advanced capabilities tailored to meet evolving business demands.

A prominent use case of Azure Stack HCI is hosting Azure Virtual Desktop (AVD) session hosts on-premises, offering organizations a secure and high-performance solution for managing hybrid workplace environments. With Azure Arc-enabled services, businesses can unify their multicloud strategies while maintaining full control over sensitive data. The latest Azure Stack HCI version 23H2 further enhances its capabilities with improved observability, expanded regional availability (including Southeast Asia and central India), and streamlined deployment features to optimize hybrid operations. Additionally, its deeper integration with Azure Kubernetes Service (AKS) and Azure Arc boosts workload density and simplifies hybrid application management, making it a compelling choice for private cloud scenarios.

One key point to remember here is organizations can use Azure Stack HCI without connecting to the Azure cloud. However, connecting to the cloud allows for additional features such as backup and disaster recovery services through the support of cloud backup services.

Please refer to Figure 8-2 for the solution overview of Azure Stack HCI for improved relevance. This diagram offers a comprehensive overview of an Azure solution, detailing its primary components and integrations. The

upper section emphasizes critical Azure services, including Entra ID, Site
Recovery, Backup, File Sync, Update Manager, Policy, Monitor, Key Vault,
and Microsoft Defender for Cloud, as well as additional functionalities
provided by Azure Arc. The central section centers on Azure Stack HCI,
illustrating traditional applications operating on Windows and Linux
virtual machines, Azure Virtual Desktop, Kubernetes-based applications,
and Arc-enabled services driven by AKS. The lower section describes the
Azure Stack HCI operating system, which includes Hyper-V and Storage
Spaces Direct, while the footer showcases leading solutions, validated
nodes, and integrated systems that enhance hybrid cloud capabilities.
Collectively, these components illustrate how Azure effectively integrates
diverse services to support contemporary IT infrastructure.

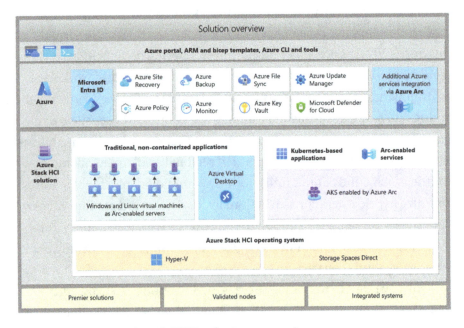

Figure 8-2. *Azure Stack HCI solution overview*

Refer to Table 8-1 to make an informed and confident decision when
selecting between Hyper-V HCI and Azure Stack HCI.

Table 8-1. *Hyper-V HCI and Azure Stack HCI*

Feature	Hyper-V in HCI	Azure Stack HCI
Deployment Model	Primarily on-premises with a focus on local infrastructure.	Hybrid by design; integrates tightly with Azure services while supporting on-premises deployments.
Cloud Connectivity	Optional integration with cloud services for backup and disaster recovery.	Can function standalone, but leveraging Azure unlocks advanced capabilities like Azure Monitor and Backup.
Management Tools	Managed through Hyper-V Manager and system center virtual machine manager (SCVMM).	Centralized management using Windows Admin Center and Azure Portal for hybrid environments.
Scale and Flexibility	Scales horizontally within the physical data center.	Supports hybrid scalability across Azure regions and on-premises infrastructure.
Software Updates	Manual or semi-automated updates depending on the tooling.	Unified Update Manager for seamless updates across hybrid and local workloads.
Backup and Recovery	Standard backup options and replicas through on-prem solutions.	Enhanced through Azure Site Recovery and Backup for hybrid resilience and disaster recovery.
Workload Optimization	Suitable for traditional virtualization workloads.	Optimized for hybrid workloads, including AI, machine learning, and virtual desktop infrastructure (VDI).
Hardware Integration	Works with a wide range of hardware, requiring some manual configurations for HCI setups.	Certified by Microsoft with validated nodes ensuring seamless integration and performance optimization.

(continued)

Table 8-1. (*continued*)

Feature	Hyper-V in HCI	Azure Stack HCI
Advanced Features	Limited advanced hybrid capabilities.	Includes cloud-integrated features like Azure Arc, Azure Policy, and Defender for Cloud.
Cost Structure	Typically involves upfront CAPEX for infrastructure.	Flexible OPEX-driven pricing model with "pay-as-you-use" capabilities for Azure services.

8.2 Real-World Use Cases for Azure HCI

As per Microsoft Industry clouds and Microsoft data center modernization guide, Azure Stack HCI has been successfully adopted across multiple sectors to tackle specific IT challenges and needs. For example, Bradley Legal Services, a law firm, implemented Azure Stack HCI to upgrade its infrastructure, facilitating the secure virtualization of sensitive legal applications. The Shielded VM feature played a crucial role in maintaining data confidentiality while ensuring compliance with regulatory standards. In a similar vein, Hendrick Motorsports, a prominent player in the auto racing industry, utilized Azure Stack HCI to boost the performance of its data-heavy applications. By incorporating Intel Optane persistent memory, they achieved swift data access, which was essential for performance analytics and decision-making on race day.

Azure Stack HCI provides a versatile and scalable option for hybrid cloud settings. It can begin with a single-node setup, making it suitable for smaller operations or branch locations. As requirements expand, it can scale to a 16-node cluster, guaranteeing high availability and performance for more extensive enterprise workloads. In the healthcare domain, Florida State College of Medicine adopted Hyper-V HCI to manage hybrid workloads with Azure Arc, allowing for efficient management of patient data across both on-premises and cloud platforms. This hybrid approach enhanced scalability while complying with stringent data governance regulations. Furthermore, Cherokee County School District employed Azure Stack HCI to implement a cost-effective disaster recovery solution. By leveraging live migration and failover clustering, they guaranteed continuous access to educational resources during hardware malfunctions.

Finally, small- and medium-sized businesses (SMBs), such as ASM Enterprise Solutions, reap the benefits of Azure Stack HCI's scalability and lower hardware expenses. Its capability to virtualize legacy applications like Active Directory and DNS while transitioning to a hybrid cloud framework has made it a favored option for modernization initiatives. Azure Stack HCI is particularly advantageous for scenarios like edge computing, which require minimal infrastructure. Its hybrid features, such as compatibility with Azure Backup and Azure Site Recovery, further strengthen its effectiveness as a solution. Let's dive into the use cases in detail to have a close look.

8.2.1 Use Case 1: Edge Computing and Branch Offices

In today's world of distributed teams and decentralized operations, edge computing has become a transformative force, with Microsoft's Azure Stack HCI at the forefront of changing how organizations oversee their remote and branch offices (ROBO). Previously restricted by physical locations and expensive, large-scale infrastructure, companies can now enjoy the freedom to extend computing power, storage, and security right to the edges of their operations. Microsoft has made a notable commitment to edge infrastructure with Azure Stack HCI, offering a compact yet robust solution tailored to the specific requirements of branch offices, remote sites, and other edge environments.

Edge computing shifts the focus of processing closer to where the data is generated, rather than relying solely on a centralized data center. This model is especially advantageous for branch offices that often face challenges related to space, cooling, noise, and installation options. Azure Stack HCI effectively tackles these issues by providing a customized, on-premises solution that supplies essential computing and storage capabilities without the burdens of traditional large-scale setups. What used to be a limitation distance from the main office has transformed into a chance for employees to carry out vital tasks from virtually anywhere, thanks to the cloud-like features of Azure Stack HCI in an edge setting.

8.2.1.1 Use Case: Retail Chain Enhances Edge Operations with Azure Stack HCI

A prominent global retail chain, with numerous stores in both urban and remote areas, encountered considerable difficulties in efficiently managing its IT infrastructure across branch locations. Each store required localized data processing to support low-latency operations, including inventory management, point-of-sale systems, and customer analytics. Furthermore, the IT team sought centralized control to implement uniform security policies while dynamically scaling operations without incurring significant hardware costs.

The retail chain implemented Azure Stack HCI to upgrade its branch office infrastructure. The initial deployment involved single-node cluster in smaller locations, offering a cost-effective yet powerful solution for localized processing, but it won't ensure availability and could be a risk in case of failure. This setup allowed essential workloads to operate independently from central data centers, reducing latency and ensuring continuous operations. Larger stores or regional hubs utilized Azure Stack HCI's capability to expand to 16-node clusters, accommodating high availability and resource-intensive applications.

Integration with Azure Arc enabled centralized governance via the Azure portal, allowing IT administrators to oversee, manage, and secure all branch deployments from a single interface. Azure services such as Backup and Site Recovery provided data protection and resilience throughout the retail chain.

Key features and benefits edge-optimized performance: Localized processing facilitated low-latency operations, especially for critical applications like real-time point-of-sale systems. The compact design of single-node clusters was ideal for smaller branch offices, while larger deployments efficiently managed more complex tasks.

The capability to transition from single-node configurations to multi-node clusters provided the necessary flexibility to adapt to changing business requirements without the need for infrastructure redesign. Azure integration facilitated effortless access to essential tools such as Azure Backup, Site Recovery, and cloud-managed security.

Integrated features like Microsoft Defender for Cloud and Trusted Launch safeguarded workloads from sophisticated cyber threats, while the secure architecture maintained adherence to corporate and regulatory standards. This solution enabled the retail chain to optimize its operations, minimizing downtime and enhancing visibility across IT assets. The scalability of Azure Stack HCI, along with its hybrid cloud functionalities, delivered an effective and future-ready IT strategy for the organization.

This use case from the Microsoft community highlights that Azure Stack HCI deployment should be aligned as per organization need. For example, single-node setups are ideal for small offices constrained by space and budget. Conversely, larger implementations require multi-node clusters to guarantee high availability and efficient workload distribution. Organizations should also prioritize robust connectivity to maximize Azure integration, utilizing tools like Azure Arc for policy enforcement and Microsoft Defender for Cloud to safeguard their infrastructure. I advise businesses to perform a comprehensive assessment of their edge requirements, concentrating on application workloads, security needs, and scalability demands. The seamless hybrid connectivity provided by Azure Stack HCI can lead to substantial operational efficiencies, but this potential is realized only when supported by well-defined governance practices and sufficient training for IT personnel. This solution is especially advantageous for sectors such as retail, healthcare, and logistics, where low latency and dependable edge computing are essential for success. For additional technical insights, I encourage you to consult Microsoft's resources on Azure Stack HCI available in their Tech Community Hub.

8.2.1.2 Simplifying ROBO Deployments with Azure Stack HCI

Azure Stack HCI is an exceptional choice for remote office and branch office (ROBO) scenarios, providing scalability, performance, and cost-effectiveness while maintaining reliability. Tailored to meet the unique demands of ROBO environments, it features compact configurations that reduce both physical space and operational costs. For instance, a two-node Azure Stack HCI setup not only lowers hardware expenses but also eliminates the requirement for high-speed switches by utilizing direct, back-to-back networking. This streamlined method is especially advantageous for smaller branch offices, retail outlets, and manufacturing facilities, where space and budget limitations often influence infrastructure decisions.

This comprehensive solution also removes the need for external storage, simplifying deployment and further cutting costs. By catering to specific environmental considerations such as noise, cooling, and form factor, Azure Stack HCI enables branch offices and edge locations to utilize on-premises resources and security essential for demanding workloads. Whether managing virtual machines (VMs) for inventory control in a retail environment or analyzing real-time data in a manufacturing setting, this configuration guarantees consistent performance, no matter the location.

For organizations with multiple branches, the savings at each location can add up significantly over time. The compact design and economical subscription model of Azure Stack HCI empower businesses to expand their infrastructure in a cost-effective manner while ensuring high availability and resilience through features like nested resiliency.

The subscription model of Azure Stack HCI functions on a subscription basis, allowing users to pay according to their resource consumption. This model offers businesses the flexibility to align their expenses with actual usage, enabling them to modify their capacity as required, which is especially beneficial for scaling workloads in a hybrid cloud setting.

In terms of pricing strategy, there has been a notable transition toward operational expenditure (OPEX) models, which can be beneficial for organizations looking to minimize initial capital outlays. Recent updates have eliminated specific fees, such as the $10 per physical core per month charge for Azure Stack HCI, along with other costs associated with Windows Server virtualization and Azure Kubernetes Service, which are now waived for qualifying customers. This adjustment makes the platform more cost-effective and manageable. Furthermore, Azure Stack HCI is frequently offered in conjunction with Azure Support subscriptions, aiding in the simplification of management expenses.

Additionally, businesses can take advantage of Azure Stack HCI's hybrid capabilities, seamlessly integrating on-premises infrastructure with Azure cloud services, which facilitates a cohesive and scalable method for workload management. The pricing flexibility of Azure Stack HCI enhances its attractiveness, making it suitable for a wide range of workloads, including SQL, Kubernetes, and Azure Virtual Desktop implementations.

Furthermore, its integration with Azure facilitates centralized monitoring and management through Azure Arc, providing IT administrators with a cohesive view of deployments across all sites from the Azure portal. This minimizes administrative workload and ensures uniform policy application.

Azure Stack HCI is designed to accommodate low-core-count servers, making it particularly suitable for Remote Office/Branch Office (ROBO) environments that operate with fewer than 12 virtual machines. By providing options such as four or eight core servers, organizations can effectively address their workload requirements without the risk of overprovisioning. Furthermore, leveraging Azure services for cluster quorum witness, backup, and security significantly enhances cost efficiency by minimizing the necessity for extra on-premises infrastructure.

Ultimately, Azure Stack HCI enables organizations to advance their ROBO or edge infrastructures, striking an ideal balance between performance, reliability, and cost-effectiveness. Its seamless integration with Azure facilitates scalability and operational ease, positioning it as a fundamental element in the modernization of data centers. Whether serving a global retail chain or a local manufacturing facility, Azure Stack HCI allows edge locations to operate with the same efficiency as central data centers, all while maintaining a streamlined, scalable, and budget-friendly approach.

8.2.1.3 Leveraging Azure Integration for Centralized Management

Azure Stack HCI's connection with the wider Azure ecosystem elevates edge computing significantly. With tools like Azure Arc, administrators can oversee all their Azure Stack HCI setups, whether located in main data centers or remote branch offices, from one cohesive interface. This centralized management approach significantly eases the burden on IT teams, enabling them to monitor and manage infrastructure via the Azure portal, no matter where the physical equipment is situated.

For branch offices, this translates to a simpler management experience, as the challenges of overseeing multiple locations are minimized. IT administrators are no longer required to be on-site for routine maintenance or troubleshooting tasks. Instead, they can handle these responsibilities from a central office or even remotely, which greatly cuts down on the time and resources needed to keep edge infrastructure running smoothly. Additionally, Azure integration empowers businesses to utilize cloud services like backup, security, and cluster quorum witness, all of which can be managed off-site through the Azure cloud. This not only reduces the need for extra local infrastructure but also helps lower costs and streamline the management process.

8.2.2 Use Case 2: Disaster Recovery and Business Continuity

A meticulously designed disaster recovery (DR) plan guarantees the swift restoration of vital systems and data, significantly reducing operational interruptions. In contrast, business continuity (BC) emphasizes the preservation of essential operations during and after a disaster, ensuring that organizations can continue to provide critical services with minimal disruption. These approaches work in tandem, not merely for survival but for fostering resilience, empowering businesses to adapt and flourish amidst unexpected challenges.

In essence, establishing a robust DR and BC framework represents a strategic investment in the stability of an organization, assuring that both employees and customers can depend on consistent service availability, even in times of crisis. This framework transcends being a mere safety measure; it is a crucial strategy for promoting sustained growth and maintaining a competitive edge in the marketplace.

Figure 8-3 illustrates how to design and implement disaster recovery of Azure Stack HCI by using stretched clustering.

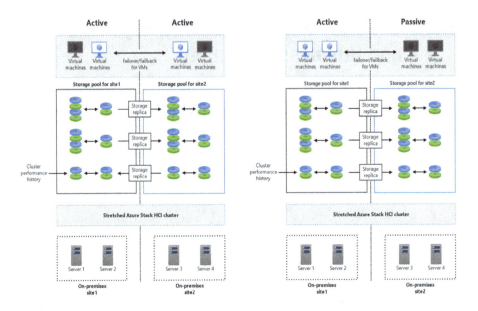

Figure 8-3. *Stretched cluster architecture of Azure Stack HCI*

This illustration presents the stretched cluster architecture of Azure Stack HCI, which provides exceptional disaster recovery (DR) solutions tailored for enterprise settings. It features two distinct configurations: Active-Active and Active-Passive, showcasing the adaptability in workload distribution and the management of failover processes across various on-premises locations. An in-depth analysis of the essential components is provided to assist you in determining whether Azure Stack HCI is a suitable fit for your organization's DR strategy.

8.2.2.1 Stretched Cluster with Active-Active Configuration

An active-active configuration involves both locations concurrently managing virtual machines (VMs) and handling workloads. The accompanying diagram demonstrates the synchronization of storage pools between Site 1 and Site 2 through Storage Replica, a functionality that enables seamless data replication across sites to enhance redundancy. This setup guarantees high availability and minimal downtime, allowing workloads to switch over or revert to the alternate site in case of a site failure. The active-active architecture is particularly beneficial for organizations that need to distribute loads across different locations, including multinational companies or financial institutions that operate latency-sensitive applications.

8.2.2.2 Stretched Cluster with Active-Passive Configuration

In an active-passive configuration, Site 1 is responsible for managing all active workloads, while Site 2 remains on standby for disaster recovery. This arrangement minimizes the operational burden on the passive site, yet guarantees that essential data is consistently synchronized through Storage Replica. In the event of a failover, workloads can be smoothly shifted to the passive site, thereby maintaining business continuity. This model is especially advantageous for sectors that face stringent budget limitations but have significant disaster recovery needs, such as retail chains and manufacturing facilities.

Key advantages of organizational disaster recovery planning:

Figure 8-4 indicates the DR planning benefits with Azure Stack HCI, which you can leverage for your organization's DR plan.

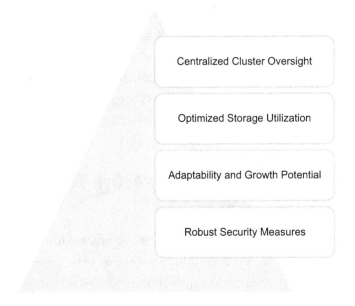

Figure 8-4. *Key advantages of DR planning with Azure Stack HCI*

Centralized cluster oversight: Both setups take advantage of Azure Stack HCI's seamless integration with Azure Arc, allowing administrators to oversee performance and replication status through a unified interface within the Azure portal.

Optimized storage utilization: By employing local storage pools and replication rather than relying on centralized external storage, the stretched cluster reduces hardware dependencies, leading to cost savings while ensuring reliability.

Adaptability and growth potential: The architecture accommodates scaling from small two-node configurations to larger clusters, providing versatility suitable for a range of business sizes and workloads.

Robust security measures: Azure Stack HCI features integrated encryption and Trusted Launch capabilities to protect sensitive data both in transit and at rest, which is essential for meeting compliance requirements in regulated sectors.

In summary, Azure Stack HCI's stretched cluster solution empowers organizations to choose a DR plan that balances cost, complexity, and performance. Whether prioritizing real-time operations with an active-active model or focusing on cost efficiency with an active-passive configuration, this architecture ensures that your critical workloads remain resilient and uninterrupted during disasters.

8.2.2.3 Financial Services: Securing Critical Transactions

Picture the turmoil that ensues when a financial institution experiences an outage during peak trading hours. Every moment counts, and any downtime can lead to substantial financial losses and harm the institution's reputation. For banks and financial organizations, Hyper-V in HCI offers the essential replication features to safeguard vital financial data and transactions. Virtual machines (VMs) that support trading platforms, online banking, and payment processing are replicated synchronously in real time, ensuring that if a primary system fails, a backup is instantly available to take over.

Additionally, Hyper-V's seamless integration with Azure Site Recovery allows for automated failover across different geographic locations. This capability ensures that financial activities, whether it's real-time trading or handling large volumes of transactions, can proceed without any interruptions. For institutions that must adhere to strict compliance standards like PCI DSS or SOX, Hyper-V guarantees that data remains secure and accessible, even during disasters. The option to switch to a secondary data center or the cloud provides protection for both the institution and its clients against the fallout from an outage.

8.2.2.4 Healthcare: Preserving Patient Care in Crisis

In the healthcare sector, downtime isn't just a minor setback; it can pose serious risks to patient safety. Hospitals and healthcare providers depend on constant access to essential information like patient records, lab results, and clinical systems. Hyper-V in HCI plays a crucial role in keeping these vital systems running smoothly, even when hardware failures or unforeseen problems arise. For healthcare organizations, the high availability and disaster recovery capabilities of Hyper-V are essential for safeguarding sensitive information while ensuring continuous patient care.

Imagine a situation where a hospital's main server, which supports its electronic medical records (EMR) system, goes down. Thanks to Hyper-V's failover clustering and real-time replication, the EMR system can swiftly shift to a secondary node or an off-site location, allowing healthcare professionals to access the necessary patient data without any interruptions. This smooth transition is especially important during emergencies when timely patient care is critical. Additionally, healthcare organizations can enhance their disaster recovery plans by integrating Azure Site Recovery, providing an extra layer of protection against even the most significant disasters.

8.2.2.5 Retail: Keeping Sales Systems Running

In the retail industry, any downtime can significantly affect customer satisfaction, result in lost sales, and harm brand loyalty. Retailers depend on point-of-sale (POS) systems, inventory management tools, and online services to operate seamlessly, whether during busy holiday seasons or regular hours. Hyper-V in HCI provides an ideal solution for retail businesses aiming to maintain continuity across various store locations.

For instance, a large retail chain might utilize Hyper-V's replication capabilities to duplicate their POS and inventory systems across multiple data centers or even to the cloud. If the main system at a flagship store encounters an issue, the replicated system activates, allowing transactions to proceed without interruption. Additionally, the retail chain can leverage Hyper-V's integration with Azure to safeguard essential data like sales reports and inventory counts, minimizing the risk of data loss and shielding the business from potential revenue setbacks.

8.2.2.6 Manufacturing: Keeping the Production Line Moving

In the manufacturing sector, even a brief period of downtime can result in significant production delays and missed deadlines. For factories that depend on real-time data from their machines and production systems, Hyper-V in HCI offers a strong disaster recovery solution that keeps operations running smoothly. By implementing replication and automated failover, manufacturers can protect vital production systems from interruptions, whether they stem from power outages or equipment failures.

Consider a manufacturing facility situated in a remote area. If the primary system managing the assembly line encounters a failure, production would usually halt, causing order delays and financial repercussions. However, with Hyper-V's low-latency replication, the system can swiftly transition to a secondary node or an off-site backup, allowing production to proceed without interruption. Additionally, manufacturers can take advantage of Azure's cloud services to back up critical data and systems, ensuring rapid recovery in case of a disaster. This hybrid disaster recovery approach provides both flexibility and cost efficiency by eliminating the necessity for a fully equipped secondary data center.

8.2.2.7 Education: Safeguarding Learning Resources

Educational institutions, particularly universities and online learning platforms, must maintain uninterrupted access to their learning management systems (LMS). Both students and faculty depend on reliable access to digital classrooms, assignments, and communication tools. Hyper-V in HCI offers a robust disaster recovery (DR) solution for these institutions, allowing classes to proceed smoothly even during system failures.

For example, during critical exam periods, a disruption in the LMS could hinder students from accessing essential study materials or submitting their work. By utilizing Hyper-V's automated failover to a DR site, schools can keep their systems operational. Additionally, with Azure integration, educational institutions can back up their virtual environments and resources in the cloud, safeguarding both data and services.

8.2.2.8 Conclusion: Hyper-V in HCI – The Backbone of Modern Disaster Recovery

In various contexts, be it a bank securing transactions, a healthcare facility ensuring patient safety, a retailer maintaining seamless sales, or a manufacturer keeping production on track Hyper-V in HCI serves as a fundamental element of a comprehensive disaster recovery and business continuity plan. Its features, including real-time data replication, automated failover processes, and seamless integration with Azure, position Hyper-V in HCI as an essential resource for organizations aiming to protect their operations against unexpected interruptions.

In an environment where even brief periods of downtime can lead to substantial financial, operational, or human costs, the capacity for rapid recovery from disasters has become imperative. Hyper-V in HCI provides a cost-efficient, scalable, and dependable solution that empowers businesses to maintain their operations, even in the face of adversity. While the future remains uncertain, organizations can trust that with Hyper-V in HCI, they are well-prepared to tackle any challenges that may arise.

8.2.3 Use Case 3: High-Performance Virtual Desktop Infrastructure (VDI) – Empowering Remote Workforces

Azure Virtual Desktop (AVD) on Azure Stack HCI is tailored to tackle significant issues that organizations encounter when handling contemporary virtual desktop environments. By merging Azure's powerful VDI capabilities with the on-premises framework of Azure Stack HCI, it effectively resolves various challenges while delivering a versatile and high-performance hybrid solution.

A significant issue that AVD on Azure Stack HCI addresses is the latency encountered when accessing on-premises applications. In cases where essential applications, such as enterprise resource planning (ERP) systems or databases, are located locally, conventional cloud-based virtual desktops often fail to provide the low-latency performance that users require. AVD on Azure Stack HCI effectively tackles this problem by allowing organizations to position session hosts in proximity to these vital applications. This strategic placement minimizes the latency between virtual desktops and on-premises applications, leading to an enhanced user experience and improved application performance, particularly for workloads that are sensitive to latency. A typical architectural setup for Azure Virtual Desktop is illustrated in the following diagram:

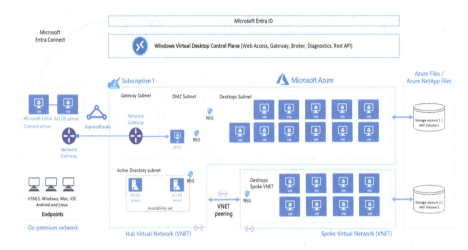

Figure 8-5. *Architectural setup for Azure Virtual Desktop*

The complexity of managing infrastructure continues to grow, particularly when navigating various vendors, credentials, and protocols. Azure Stack HCI provides a cohesive virtualization solution that integrates effortlessly with Azure services, making this environment more manageable. With AVD on Azure Stack HCI, organizations can oversee clusters and workloads from the Azure portal, removing the necessity for multiple management tools and unifying operations within a single, user-friendly interface. For organizations already leveraging Azure services, this integrated management approach significantly diminishes operational challenges and enhances the efficiency of their IT processes.

When performance is critical, particularly in edge situations, the ability to position Azure Stack HCI clusters closer to critical applications is essential. For example, implementing AVD on a cluster situated close to an ERP system guarantees reduced latency and enhanced responsiveness. This flexibility enables organizations to customize their infrastructure to address particular workload needs, whether in remote offices, manufacturing sites, or branch locations. By strategically aligning infrastructure deployment with the proximity of workloads, companies can provide an exceptional user experience while adhering to rigorous performance standards.

The smooth incorporation of AVD with Azure Stack HCI enhances the virtual desktop experience significantly. Capabilities like Windows 10/11 multi-session enable organizations to optimize resource use while maintaining high performance levels. The architecture is designed to leverage advanced optimizations, including FSLogix, which facilitates quick user logins and effective profile management. When paired with Azure Active Directory for secure authentication and conditional access, users benefit from a swift, secure, and efficient desktop environment customized to their specific requirements.

Azure Virtual Desktop on Azure Stack HCI is a crucial solution for organizations looking to harmonize the advantages of cloud scalability with the performance requirements of their on-premises systems. This hybrid architecture effectively tackles issues such as latency, complexity, and user experience, enabling businesses to attain operational excellence while enhancing both cost efficiency and productivity. Let's explore some use cases of AVD for better understanding, taking different domains into consideration.

8.2.3.1 The Financial Industry: Securing Remote Access

Imagine a large financial organization that needed to swiftly implement secure remote access for its staff during the worldwide transition to working from home. This organization, which had to adhere to stringent data protection regulations, faced the task of keeping sensitive financial information safe while allowing remote employees to work effectively. By utilizing Hyper-V in a hyper-converged infrastructure (HCI), the organization successfully established a high-performance virtual desktop infrastructure (VDI) that granted employees secure access to their desktops, applications, and data within a well-managed virtual environment.

Thanks to Hyper-V's scalability and the strong foundation of HCI, the IT team was able to quickly create virtual desktops without making substantial hardware investments. This flexibility enabled the organization to expand its VDI solution as needed, accommodating more remote workers seamlessly. Additionally, by integrating with Azure Active Directory and Azure Multi-Factor Authentication (MFA), the organization ensured that all remote connections were secure, providing reassurance that vital financial data remained protected, even while employees worked from home.

8.2.3.2 Healthcare: Providing Remote Access to Medical Staff

In the healthcare industry, where having access to patient information is essential for providing top-notch care, VDI powered by Hyper-V in HCI has emerged as a key solution for facilitating secure remote access for healthcare professionals. For example, a large hospital network faced the challenge of allowing doctors, nurses, and administrative personnel to access their workstations and medical records from home or remote clinics. By utilizing Hyper-V in HCI, the hospital successfully established a secure VDI environment that enabled medical staff to remotely access patient data while adhering to HIPAA and other healthcare regulations.

By deploying their VDI setup on Hyper-V within an HCI framework, the hospital enjoyed benefits like high availability and fault tolerance. This meant that even if there were hardware malfunctions or network disruptions, medical personnel could still reach vital systems. The capability to replicate virtual desktops across various nodes added an extra layer of dependability, ensuring uninterrupted access to patient records, which is particularly crucial during emergencies or natural disasters when remote access is even more vital.

8.2.3.3 Education: Supporting Hybrid Learning

As educational institutions transitioned to hybrid learning models, ensuring that students and faculty had dependable access to virtual desktops became crucial for a smooth learning experience. Schools and universities adopted Hyper-V in HCI to implement VDI solutions that catered to both on-campus and remote learners. For instance, one university utilized Hyper-V in HCI to establish virtual computer labs, allowing students to use demanding applications like AutoCAD, MATLAB, and other resource-heavy software from their personal devices, whether they were on campus or studying from home.

Thanks to Hyper-V's effective resource management and HCI's scalable capabilities, the university successfully supported thousands of simultaneous virtual desktops, granting students and faculty consistent access to essential tools. This VDI solution also streamlined IT management, enabling the university's IT team to deploy, oversee, and update virtual desktops from a central location, which simplified the process of managing individual physical desktops throughout the campus.

8.2.3.4 Manufacturing: Enabling Remote Monitoring and Collaboration

In the world of manufacturing, the emergence of Industry 4.0 and smart factories has opened up exciting possibilities for remote monitoring and teamwork among global teams. For a major manufacturing firm, utilizing Hyper-V in a hyper-converged infrastructure (HCI) was instrumental in allowing their engineers and technicians to connect to their workstations and production systems from afar. By implementing a virtual desktop infrastructure (VDI) on Hyper-V, the company's engineers could oversee production lines, resolve machinery issues, and collaborate on product designs from any location worldwide.

This capability for remote access not only boosted productivity but also lessened the necessity for engineers to be physically present on the factory floor, which was especially important during travel restrictions and social distancing measures. Hyper-V's strong networking and virtualization features provided a seamless, high-performance experience for users, even when they were using demanding applications like CAD and simulation software remotely. The ability to replicate VDI sessions across various nodes ensured high availability, reducing the risk of disruptions to essential manufacturing processes.

8.2.3.5 Retail: Supporting Distributed Workforce

Retailers, especially those with teams spread across different locations, depend significantly on remote workers for various tasks like managing inventory, coordinating the supply chain, and providing customer service. For a global retail chain, utilizing Hyper-V in HCI was the perfect solution for implementing a VDI system that enabled remote access for employees situated in multiple areas. Thanks to Hyper-V's scalability, the retail chain could effortlessly manage seasonal increases in remote staff, particularly during busy times like the holiday shopping season, when extra personnel are needed to handle the surge in sales and customer inquiries.

By operating their VDI setup on Hyper-V, the retail chain enjoyed the advantages of centralized management and automated updates, which empowered their IT team to swiftly deploy new virtual desktops or refresh existing ones without needing to visit each site physically. Additionally, the integration with Azure offered further flexibility, allowing the company to keep backups and disaster recovery copies of their virtual desktops in the cloud, thus ensuring business continuity in the event of system failures.

8.2.3.6 Conclusion: Hyper-V in HCI – Revolutionizing Remote Work with VDI

Hyper-V in HCI offers organizations an ideal platform for implementing high-performance VDI solutions. Sectors such as finance, healthcare, education, manufacturing, and retail are finding that the ability to provide secure, scalable, and efficient virtual desktops is a significant advantage for empowering their remote teams. The synergy of Hyper-V's virtualization capabilities, HCI's scalability, and Azure integration allows organizations not just to adapt to remote work but to excel in this new landscape. Whether it's safeguarding sensitive information, ensuring consistent availability, or delivering a smooth user experience, Hyper-V in HCI is assisting various industries in meeting the challenges of contemporary work settings, all while cutting costs and streamlining IT management. As remote work continues to progress, Hyper-V in HCI stands out as a leader in offering secure, dependable, and high-performance VDI solutions that contribute to business success.

8.2.4 Use Case 4: Architecting for Data-Centric Workloads with a Big Data Focus

Azure Stack HCI has progressed to meet the demands of contemporary data-driven workloads, establishing itself as an attractive platform for big data and analytics. With the incorporation of significant enhancements such as Azure Arc-enabled infrastructure and Lifecycle Manager (launched in version 23H2), it facilitates a cohesive experience for the management, deployment, and orchestration of intricate environments vital for big data activities.

Azure Stack HCI transcends traditional infrastructure solutions, serving as a pivotal platform for unlocking the full potential of big data and analytics. At its foundation, it revolutionizes data performance through low-latency and high-throughput capabilities, enabling the seamless ingestion and processing of extensive datasets. It supports distributed systems such as Apache Hadoop and Spark, facilitating both batch and real-time workflows, thereby equipping organizations to address data challenges with unparalleled speed and efficiency.

What distinguishes Azure Stack HCI is its seamless integration with Azure, allowing businesses to harness cloud scalability while retaining local control over sensitive information. The hybrid management capabilities provided by Azure Arc support Kubernetes clusters, expansive data lakes, and advanced machine learning workflows.

The addition of the Lifecycle Manager significantly boosts operational efficiency by streamlining updates across both hardware and software layers. With automated updates for the operating system, drivers, and core services, it guarantees high availability essential for analytics-driven environments.

Furthermore, Azure Stack HCI excels in real-time and predictive analytics. IoT devices, real-time messaging platforms like Kafka, and Azure Event Hubs collaborate effectively to deliver immediate insights. For predictive analytics, Azure Synapse Analytics offers sophisticated forecasting tools, converting raw data into actionable intelligence.

Security and compliance are paramount, featuring encrypted storage and stringent access controls that satisfy the rigorous requirements of sectors such as healthcare and finance. Combined with its elastic scalability – enabled by stretch clustering and GPU-accelerated computing – Azure Stack HCI serves as a robust solution for dynamic workloads.

Ultimately, Azure Stack HCI is more than mere infrastructure; it is a designed solution that reimagines how businesses extract value from their data in a hybrid environment. Engage with it to revolutionize your organization's approach to big data. It integrates on-premises and cloud-based analytics, catering to organizations that prioritize data sovereignty without sacrificing analytical capabilities.

8.2.4.1 Healthcare: Enhancing Patient Outcomes with Data Analytics

Imagine a vast healthcare network that processes petabytes of patient information daily. This includes everything from medical images and patient records to lab results and genomic data. Healthcare organizations bear a significant responsibility not just to store this information but also to analyze it to enhance patient outcomes. By utilizing Hyper-V in a hyper-converged infrastructure (HCI), this healthcare network has established a cohesive platform that facilitates data-driven analytics, leading to quicker and more precise diagnoses.

With Hyper-V's virtualization features integrated into the HCI setup, the organization can effortlessly scale its computing resources to meet the growing demands for data processing, particularly for high-resolution medical imaging and sophisticated analytics. For example, AI-powered imaging analysis can swiftly identify early indicators of diseases, like cancer, much more effectively than conventional techniques. This minimizes diagnostic errors and equips physicians with vital insights to elevate patient care. Additionally, the collaboration with Azure AI enables the healthcare network to implement machine learning algorithms on patient data, allowing them to uncover patterns that enhance treatment strategies and better predict patient outcomes.

8.2.4.2 Retail: Powering Predictive Analytics for Consumer Behavior

In the retail industry, grasping customer behavior is essential for maintaining a competitive edge. Retailers collect extensive data on customer purchases, preferences, and trends, but turning this data into actionable insights necessitates a robust infrastructure capable of executing real-time analytics. Hyper-V in HCI provides retail businesses with the flexibility and performance required to conduct predictive analytics, enhancing inventory management, sales strategies, and customer engagement.

For instance, a global retailer leveraged Hyper-V in HCI to create a scalable analytics platform that could handle terabytes of data from their physical stores, online channels, and loyalty programs. By virtualizing their analytics tasks, they gained the ability to analyze customer purchasing trends in real time, allowing them to forecast product demand across various regions. This optimization of inventory levels ensured that popular items were consistently available while minimizing waste from overstocked goods. Additionally, the retailer utilized Azure Data Lake for the long-term storage of historical data, which facilitated deeper analyses and informed data-driven decisions that improved customer experiences and boosted revenue growth.

8.2.4.3 Financial Services: Accelerating Fraud Detection and Risk Management

In the financial services sector, leveraging data-driven workloads is essential for detecting fraud, managing risks, and ensuring compliance with regulations. Financial institutions handle vast quantities of data every moment, encompassing everything from transactions and customer information to market dynamics and investment strategies. Hyper-V in HCI offers these organizations a robust platform capable of meeting the heavy computational needs of big data processing while maintaining data security and adhering to regulations such as GDPR and PCI DSS.

For example, a major bank implemented Hyper-V in HCI to create a real-time fraud detection system. By employing machine learning algorithms to scrutinize transaction data as it occurred, the bank was able to spot suspicious activities and flag potentially fraudulent transactions before they were finalized. The low-latency networking features of Hyper-V allowed the bank to process these transactions swiftly and effectively, reducing the risk of financial losses. Furthermore, the bank utilized Azure

Synapse Analytics to consolidate data from multiple sources, including transaction records, customer profiles, and third-party risk assessments, thereby establishing a comprehensive risk management strategy that enhanced their ability to combat fraud and improve compliance with regulations.

8.2.4.4 Manufacturing: Optimizing Production with IoT and Big Data

Manufacturing companies are increasingly turning to IoT (Internet of Things) devices to collect data from their production lines, machinery, and supply chains. These data streams offer crucial insights into equipment performance, energy consumption, and production efficiency. However, handling and analyzing such vast amounts of data necessitates a strong infrastructure. Hyper-V in HCI provides manufacturers with an excellent platform to manage their IoT workloads, process data in real time, and enhance their operations.

For instance, a leading automotive manufacturer implemented Hyper-V in HCI to create an IoT-driven analytics platform that monitored their assembly line machinery in real time. By virtualizing their data processing tasks, they were able to collect information from thousands of sensors integrated into their machinery, tracking various metrics such as temperature, vibration levels, production output, and equipment wear. This data was analyzed using Azure Machine Learning, enabling the company to spot patterns that signaled potential machine failures or inefficiencies in the production workflow.

By identifying these issues early on, the manufacturer could schedule predictive maintenance, which minimized downtime and prolonged the lifespan of their equipment. Furthermore, real-time data analysis allowed them to optimize energy consumption and reduce waste, resulting in significant cost savings and a more sustainable production process.

8.2.4.5 Energy: Driving Efficiency in Resource Management

Energy companies, especially those focused on renewable sources, depend heavily on data to manage their resources efficiently. From tracking wind turbines and solar panels to monitoring energy usage across a grid, these organizations produce enormous datasets that must be processed and analyzed swiftly to enhance performance and facilitate real-time decision-making. Hyper-V in HCI allows energy companies to manage these data-intensive tasks effortlessly, delivering the computational strength required to analyze information from various sources in real time.

For instance, a renewable energy firm utilized Hyper-V in HCI to create a data analytics platform that oversaw wind farms in multiple locations. By analyzing data from wind turbines in real time, the company was able to modify turbine settings to maximize energy production based on prevailing weather conditions, wind speed, and equipment efficiency. The scalability of Hyper-V enabled the firm to easily broaden their analytics capabilities as they incorporated additional wind farms into their operations, while Azure's cloud integration provided the means to store extensive historical data for future analysis and reporting.

8.2.5 Use Case 5: Empowering On-Premises Container and Kubernetes Workloads

The swift integration of containers and Kubernetes has revolutionized the landscape of application development and deployment, creating a demand for strong on-premises solutions. For organizations aiming to achieve cloud-like agility while tackling issues such as data sovereignty, low-latency needs, and compliance with industry standards, empowering these workloads locally is vital. On-premises Kubernetes platforms, like Azure Kubernetes Service (AKS) on Azure Stack HCI, deliver the cloud-native benefits of containerization, such as scalability, resilience, and easier management, directly within the organization's data centers.

This strategy is particularly important for sectors such as healthcare, finance, and manufacturing, where the ability to process data in real time, maintain secure application environments, and operate locally is essential. By deploying Kubernetes on-premises, organizations can leverage microservices, adopt DevOps methodologies, and facilitate edge computing workloads while retaining full control over their infrastructure. Furthermore, hybrid capabilities allow for smooth integration with public cloud services, enhancing disaster recovery, scalability, and advanced analytics, thus providing a balanced approach.

Enabling on-premises Kubernetes workloads is not merely a strategic choice; it is a critical requirement for enterprises looking to modernize their applications, enhance operational efficiency, and secure long-term IT agility.

Azure Stack HCI, when integrated with Azure Kubernetes Service (AKS) on-premises, represents a groundbreaking solution for contemporary IT landscapes. The following architecture facilitates the deployment of containerized workloads with enhanced availability and streamlined automation, effectively connecting on-premises infrastructure with cloud advancements and opening up unparalleled opportunities for businesses.

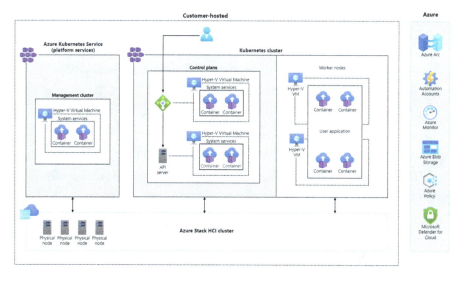

Figure 8-6. *Kubernetes cluster on Azure Stack HCI*

8.2.5.1 Revolutionizing Healthcare: Enabling Data-Driven Patient Care

Envision a healthcare landscape where critical decisions are informed by real-time insights extracted from extensive patient data. Healthcare organizations are tasked with the dual responsibility of securely managing vast quantities of sensitive information, such as electronic health records (EHRs), high-resolution medical imaging, and genomic datasets, while also utilizing this data for actionable insights. Azure Stack HCI, in conjunction with Kubernetes and containerized workloads, presents a groundbreaking solution in this field, merging state-of-the-art infrastructure with sophisticated analytics to enhance patient outcomes.

By utilizing Azure Kubernetes Service (AKS) on Azure Stack HCI, healthcare providers can efficiently deploy and manage scalable containerized applications tailored for data-intensive operations. For instance, AI-powered image recognition models running on AKS can evaluate medical scans in real time, detecting early indicators of diseases such as cancer. This capability significantly improves diagnostic precision and speed, minimizes human error, and facilitates timely medical interventions. The hyper-converged infrastructure of Azure Stack HCI offers low-latency performance, which is crucial for executing compute-intensive tasks like genomic sequencing and advanced predictive analytics.

Moreover, the hybrid capabilities of Azure Stack HCI broaden the scope of data analytics in healthcare. Sensitive patient information can be retained on-premises to adhere to stringent regulatory requirements like HIPAA while still securely connecting to Azure cloud services for AI-enhanced analysis and long-term data storage. The integration of Azure AI and machine learning tools within this framework can reveal essential patterns in patient data, paving the way for personalized treatment approaches and predictive care models that not only save lives but also optimize resource utilization.

The platform's integration with Kubernetes significantly enhances the capabilities of DevOps teams in the healthcare sector, allowing for swift innovation and the seamless deployment of new diagnostic applications and updates without disrupting services. This technology supports telemedicine initiatives and guarantees continuous access to patient records across various hospital networks, ensuring high availability, scalability, and operational resilience in contemporary healthcare settings. Combining on-premises infrastructure with the scalability of the cloud and leveraging advanced AI, Azure Stack HCI revolutionizes the way healthcare organizations process data, provide care, and adapt to the fast-changing medical environment. This architectural integration leads to improved patient outcomes, optimized operations, and positions healthcare providers as leaders in medical innovation.

8.2.5.2 Empowering Financial Service: Data Sovereignty

Azure Stack HCI represents a groundbreaking solution for financial institutions aiming to upgrade their infrastructure, improve compliance, and enhance agility in a dynamic and heavily regulated environment. In the context of real-time trading systems, where every millisecond is crucial, Azure Stack HCI provides low-latency, high-performance computing capabilities specifically designed for workloads that demand high availability and robust failover mechanisms. This guarantees continuous operations even during peak trading periods. Additionally, its integration with Azure Kubernetes Service (AKS) boosts the scalability of containerized trading applications, optimizing resource distribution and ensuring smooth performance during market fluctuations.

The critical areas of risk management and fraud detection in the financial sector greatly benefit from the hybrid cloud analytics capabilities of Azure Stack HCI. By utilizing Azure Synapse Analytics and machine learning algorithms, institutions can conduct sophisticated fraud detection and predictive analysis, allowing them to identify transaction anomalies and trends while maintaining data sovereignty – an essential factor for adhering to regulatory requirements. Furthermore, legacy core banking systems, which often suffer from inefficiencies, can be revitalized through Azure Stack HCI. Its support for virtualization and containerization, along with Azure Arc, facilitates seamless management across both on-premises and cloud environments, enabling the swift deployment of new features while preserving the stability of legacy systems throughout the modernization process.

In terms of security and compliance, Azure Stack HCI features advanced capabilities such as virtualization-based security (VBS) and encrypted storage, safeguarding sensitive financial information. Its capacity to locally host data while integrating with Azure's compliance management tools makes it an optimal choice for institutions operating across various jurisdictions with strict regulatory frameworks.

Additionally, transitioning from outdated infrastructure to Azure Stack HCI's hyper-converged solutions allows financial institutions to greatly minimize their data center size and operational expenses. Capabilities like stretch clustering and scale-out storage provide enhanced availability and scalability, while centralized management tools streamline administration and reduce costs.

By integrating on-premises infrastructure with Azure cloud services, Azure Stack HCI enables financial institutions to maintain a competitive edge in a rapidly changing digital environment. Its combination of performance, scalability, and compliance makes it a fundamental element for modernizing financial operations and fostering innovation within the sector.

8.3 Spotlight on Microsoft's Future Advancements and Windows Server 2025

Windows Server 2025 presents a robust array of virtualization features designed to meet the varied needs of different industries. A key highlight is its enhanced hybrid cloud functionality, particularly when integrated with Azure Stack HCI. This synergy facilitates the streamlined management of both on-premises and cloud resources through a single interface, making it an excellent choice for organizations moving toward a hybrid cloud infrastructure. For enterprises utilizing virtual machines (VMs), the addition of GPU partitioning support for select Nvidia models significantly boosts performance for AI applications, remote desktop usage, and high-performance computing tasks. The virtualization capabilities of Windows Server 2025 enable effective distribution of GPU resources, allowing multiple VMs to share these resources, thus providing a cost-efficient solution for demanding computational requirements without the necessity for dedicated hardware for each VM.

Moreover, the incorporation of hot patching, a feature adapted from Azure, enhances the patching process by permitting security updates to be applied to active servers without the need for restarts. This is especially beneficial for organizations with critical workloads that require minimal downtime.

In terms of storage, Windows Server 2025 supports the Resilient File System (ReFS), which enhances data integrity by safeguarding against rollback attacks and optimizing performance. The system also introduces sophisticated deduplication and compression features, which can significantly lower storage costs in virtualized settings.

For scenarios involving disaster recovery or high-availability setups, Windows Server 2025 offers enhanced failover clustering capabilities, including campus clusters and support for multi-site configurations, ensuring that essential workloads remain robust, even across different geographical locations. The updated version enhances networking capabilities via software-defined networking (SDN), introducing features such as tag-based network security groups to improve workload isolation, particularly in scenarios that necessitate the flexible relocation of virtual machines across different locations.

In conclusion, whether addressing the needs of a large enterprise aiming to upgrade its infrastructure or a smaller organization implementing hybrid cloud approaches, Windows Server 2025 offers adaptable and scalable solutions that optimize virtualization, strengthen security, and increase operational efficiency. Let's see the real power of Server 2025 with a few more features in detail.

8.3.1 The Power of AI and Automation

One of the most thrilling aspects of Windows Server 2025 is its strong emphasis on artificial intelligence (AI) and automation. Microsoft is making significant strides in AI, revolutionizing the way servers are managed, maintained, and optimized. With AI-powered insights, businesses will have the ability to foresee server issues before they arise, adjust workloads using real-time data, and automate routine maintenance tasks. This transition toward self-healing and self-optimizing systems is not merely a technological leap; it's a vital progression in an era where data is growing at an unprecedented rate and IT teams are often overwhelmed.

AI-driven automation will play a pivotal role, enabling organizations to concentrate on more strategic goals instead of getting caught up in tedious, repetitive tasks. IT administrators can utilize the machine learning capabilities integrated into Windows Server 2025 to bolster security, detect anomalies, and enhance resource management across various environments.

8.3.2 Enhanced Security and Zero Trust Architecture

In a time when cyber threats are increasingly complex, Microsoft is placing a strong emphasis on security with the upcoming Windows Server 2025. This new version will feature advanced security frameworks based on the Zero Trust model, guaranteeing that every access request is meticulously verified, regardless of its source. Enhancements to multi-factor authentication (MFA), identity protection, and privileged access management (PAM) will further strengthen organizations' defenses against potential breaches and data leaks.

Additionally, Microsoft is introducing confidential computing in Windows Server 2025, which encrypts data during processing. This innovation ensures that sensitive information is safeguarded at all times, whether it's being transmitted, stored, or actively used, making it a significant advancement for sectors like healthcare, finance, and government that manage large amounts of sensitive data.

8.3.3 Performance and Scalability: The Next Generation of Computing Power

Windows Server 2025 is designed to embrace the future of computing and storage technologies. With its support for cutting-edge advancements in quantum computing and GPU-accelerated tasks, businesses will be equipped to tackle more intricate simulations, deep learning initiatives, and extensive data analysis with improved efficiency. This version will also introduce advanced containerization features, simplifying the deployment, management, and scaling of container-based applications both on-premises and in the cloud.

In sectors like manufacturing, which depend significantly on IoT and real-time analytics, these upgrades will foster more agile production systems, quicker decision-making, and enhanced automation on the shop floor. For organizations managing data-heavy workloads such as AI training models or high-performance computing (HPC) setups, Windows Server 2025 will provide exceptional compute density, minimizing latency and maximizing throughput for critical operations.

8.3.4 Deep Integration with Azure and Multi-cloud Flexibility

One of the standout features of Windows Server 2025 is its seamless integration with Microsoft's Azure ecosystem. In our current multi-cloud landscape, businesses need the capability to effortlessly shift workloads between on-premises, edge, and cloud settings. With the addition of Azure Arc in Windows Server 2025, organizations gain remarkable control and insight into their entire infrastructure, regardless of its location. This enhanced flexibility allows IT teams to fine-tune performance, manage costs, and ensure compliance, all while harnessing the advantages of Azure's cloud services and keeping essential workloads on-site.

Additionally, Windows Server 2025 brings robust disaster recovery and backup solutions powered by Azure. By utilizing Azure's extensive global infrastructure, organizations can safeguard and replicate their on-premises workloads in the cloud, ensuring they remain operational during unforeseen outages or cyber threats. With integrated tools for managing failover and recovery, Windows Server 2025 provides a more thorough and efficient strategy for disaster preparedness.

8.3.5 Edge Computing: Extending the Intelligent Cloud

Microsoft envisions a future that transcends the conventional data center model. With the emergence of edge computing, there is a growing demand for smart, localized processing capabilities that enable real-time decision-making without depending on a central cloud. Windows Server 2025 has been specifically crafted for edge deployments, featuring lightweight configurations and native support for IoT workloads. This makes it ideal for various settings, from remote manufacturing sites to healthcare clinics in underserved areas, allowing organizations to harness computing power right where data is generated.

This focus on edge computing is further enhanced by Azure Stack HCI, which empowers organizations to create and manage hyper-converged infrastructure (HCI) solutions at the edge with the same simplicity and familiarity as traditional data centers. The incorporation of AI-driven management tools guarantees that resources are effectively allocated, monitored, and optimized for maximum performance, even in edge environments.

8.3.6 Conclusion: A Future-Proof Solution for Modern Enterprises

With the upcoming launch of Windows Server 2025, Microsoft is once again leading the way in enterprise computing. This new version is packed with forward-thinking features, including AI, automation, hybrid cloud, and edge computing, all designed to tackle the intricate challenges faced by today's businesses while also preparing for future needs. Windows Server 2025 aims to equip organizations with the essential tools to not just navigate a digital-first landscape but to excel within it.

The improvements in security, scalability, and smart management allow businesses to innovate rapidly, confident in their infrastructure's capacity to grow and adapt. Windows Server 2025 is more than just an operating system; it serves as the cornerstone for the next wave of computing. As companies prepare for the requirements of a more interconnected and data-centric world, Windows Server 2025 is poised to drive them forward with agility, strength, and resilience.

8.4 Author's Recommendation: The Road Ahead in Virtualization and IT Strategy

As we conclude this in-depth analysis of the future of virtualization, it is clear that the landscape is undergoing significant changes. The recent acquisition of VMware by Broadcom represents a pivotal moment, not only for VMware but for the broader virtualization and IT infrastructure industry. This major development prompts essential considerations for enterprises: Should they maintain their investment in VMware's products under Broadcom's leadership, or is it prudent to explore alternative solutions that may better meet their evolving business requirements? While the acquisition suggests enhanced integration of enterprise software and hardware, particularly in cloud-native settings, it also brings forth new challenges that organizations must address in their future planning.

The future of virtualization is shifting away from merely selecting among a few established solutions; it now involves navigating a swiftly evolving landscape of cloud and edge technologies. Broadcom aims to capitalize on VMware's capabilities in modernizing IT infrastructures, such as utilizing Tanzu for Kubernetes and the VMware Cloud Foundation for hybrid cloud setups, thereby accelerating the transition toward multi-cloud and edge computing environments. As businesses pursue digital transformation and modernization, many are looking to VMware's offerings to connect legacy systems with cloud-native applications. Nevertheless, the choice to remain with VMware or transition to competitors like Nutanix, Microsoft Hyper-V, or RedHat necessitates thorough strategic consideration. The recent discontinuation of VMware's perpetual licensing model in favor of subscription-based packages under Broadcom has heightened the urgency for organizations to reassess their current infrastructure and determine whether continuing with VMware aligns with their long-term objectives. In light of these developments, organizations need to evaluate both the technical and operational ramifications. This includes financial aspects, such as the ongoing expenses associated with VMware subscriptions and the potential need for hardware upgrades, as well as architectural changes that may necessitate new disaster recovery and backup strategies. The decision-making landscape is complex. Furthermore, the persistent disruptions within the partner ecosystem and the urgent requirement for IT teams to acquire new skills to utilize emerging tools are becoming increasingly critical.

Organizations now find themselves at a pivotal juncture. They can either reinforce their commitment to VMware and align with Broadcom's vision for a more cohesive, application-centric infrastructure, or they can explore alternative solutions that may provide enhanced flexibility and cost savings over time. The future of virtualization will likely blend traditional technologies with new cloud-native platforms, and success will depend on how effectively organizations adjust their strategies to address these dual challenges. In this transformative digital age, the stakes are higher than ever, compelling businesses to remain adaptable and to continually evaluate their IT strategies to maintain a competitive edge. With careful strategic planning, organizations can not only navigate these changes but also excel in a dynamic and competitive marketplace.

8.5 Lessons Learned, Best Practices, and Key Takeaways

In this book, we've delved into a range of strategies for virtualization and modernizing infrastructure, assessing options such as Nutanix's hyper-converged infrastructure (HCI), Microsoft's Azure Stack HCI, Proxmox's open-source solutions, and RedHat's enterprise-level products. Below are some key insights and best practices designed to assist technology decision-makers.

8.5.1 Flexibility Is Key

When choosing between Nutanix, Hyper-V with Azure Stack HCI, Proxmox, or RedHat, it's crucial to have the capability to adjust to evolving workloads and business needs. Today's IT environment demands agility, and solutions that facilitate rapid scaling and customization can deliver lasting benefits. Take, for example, a worldwide financial organization that started with Nutanix for its HCI setup and later transitioned to a hybrid cloud approach with Azure. This shift enabled them to easily scale up during high transaction periods in fiscal quarters, helping them reduce infrastructure expenses while ensuring optimal performance.

8.5.2 Cost Versus Control

Proprietary solutions such as VMware are known for their reliability, but they often come with significant expenses, especially when considering possible adjustments after an acquisition. On the other hand, open-source options like Proxmox or budget-friendly choices like Hyper-V can lead to considerable cost reductions while still giving businesses the necessary oversight of their infrastructure. For instance, a manufacturing firm transitioned from a VMware-focused environment to using Proxmox and Hyper-V in certain departments, achieving a 40% decrease in licensing fees while keeping control over their virtualized operations. This decision enabled them to channel the savings into product development without compromising their operational effectiveness.

8.5.3 Multi-cloud and Hybrid Strategies Are Non-Negotiable

As more tasks transition to the cloud, hybrid environments are now commonplace. Companies that utilize solutions with robust cloud integration, such as Azure Stack HCI, will be ready to meet future challenges. By combining on-premises workloads with cloud capabilities, businesses can create a more resilient and scalable environment for growth. Take, for example, a global retail chain that implemented a multi-cloud approach by integrating Azure and AWS with its on-site data center. This setup enabled smooth backup, disaster recovery, and data analysis across different regions. Their hybrid model provided the flexibility to move workloads between platforms based on cost and performance, avoiding vendor lock-in.

8.5.4 Security Must Be Baked into Every Layer

Security has shifted from being an afterthought to a fundamental component of infrastructure planning and modernization. With the rise in data breaches and ransomware attacks, businesses require platforms that incorporate security measures that adapt to emerging threats. For example, healthcare providers frequently opt for RedHat due to its stringent security protocols, which help safeguard sensitive patient information in line with HIPAA regulations. By choosing platforms with strong security frameworks, CXOs can reduce risks and prevent expensive breaches, thereby preserving trust with both customers and regulators.

8.5.5 Focus on Automation for Operational Efficiency

Automation is revolutionizing IT operations, allowing businesses to grow while minimizing human error and operational costs. Tools such as Nutanix and Hyper-V with Azure Stack HCI are integrated with AI-powered automation features that streamline infrastructure management, enable predictive maintenance, and lessen the need for manual tasks. For instance, a telecommunications firm recently implemented Nutanix Prism, automating 80% of its regular maintenance activities, which allowed its IT team to concentrate on more strategic projects. This shift toward automation enhances efficiency and gives teams the opportunity to prioritize innovation over routine maintenance.

8.5.6 Future-Proofing Requires Vendor Diversification

Relying on just one vendor can pose significant risks, especially with VMware's recent acquisition by Broadcom, which has created some uncertainty regarding pricing and support. By diversifying their vendor options and incorporating open-source solutions like Proxmox or utilizing multi-cloud platforms, businesses can protect themselves from unexpected changes in licensing, support, or pricing structures. For example, a mid-sized logistics company enhanced its IT framework by implementing Proxmox for virtualization and RedHat for container orchestration. This approach not only helped them safeguard their infrastructure against potential vendor lock-in but also allowed for smoother transitions between platforms when market conditions shifted.

8.6 Tailored Insights for Different Industries and Organizational Sizes

While the fundamental principles of virtualization and infrastructure modernization apply universally, it's essential to recognize the unique challenges faced by different industries and organization sizes. Below are additional tailored insights that decision-makers can apply when choosing the right solution for their specific environments.

8.6.1 For Large Enterprises

Nutanix and RedHat are ideal choices for large enterprises that require strong security, scalability, and effective multi-cloud management.

Nutanix's AHV hypervisor streamlines the administration of extensive infrastructures, while RedHat's collaboration with OpenShift delivers the scalability and container orchestration essential for sectors like finance, healthcare, and government. Together, these solutions offer excellent support for hybrid cloud environments, allowing large organizations to effortlessly manage both on-premises and cloud workloads without being tied to a single vendor.

8.6.2 For Small to Medium Businesses (SMBs)

Small- and medium-sized businesses (SMBs) often prioritize cost-effectiveness, making Proxmox a fantastic choice for them. Its open-source framework allows these businesses to tailor their virtual environments without incurring the hefty licensing fees associated with proprietary solutions. Meanwhile, Hyper-V combined with Azure Stack HCI offers a scalable and budget-friendly option for organizations looking to optimize their limited resources while maintaining robust performance. For instance, a retail SMB could utilize Proxmox to save on costs while also expanding through Azure Stack HCI as they open new locations.

8.6.3 For Highly Regulated Industries

RedHat Virtualization (RHV) has established itself as a preferred option for sectors such as healthcare, finance, and government, primarily due to its strong security capabilities and dedication to compliance. Its collaboration with IBM Cloud allows organizations in these highly regulated fields to adhere to stringent standards, including HIPAA for healthcare and PCI-DSS for finance. For example, a financial services firm can confidently implement RedHat Virtualization, utilizing its integrated encryption and compliance features to safeguard sensitive customer information while functioning within a hybrid cloud framework. RedHat's commitment to open standards and its enterprise-level functionalities make it especially appealing for organizations that value transparency and control in their IT infrastructure.

Conversely, Microsoft Azure Stack HCI presents a similarly robust option for regulated industries, distinguished by its close integration with Azure. Azure Stack HCI shines in hybrid environments, providing seamless connectivity with compliance tools such as Azure Policy, Microsoft Defender for Cloud, and Azure Monitor. These resources assist organizations in meeting regulatory obligations like GDPR for data protection in Europe, ISO 27001 for security, and HIPAA for healthcare. Furthermore, Microsoft's emphasis on automation streamlines the process of compliance, alleviating the administrative load on IT departments.

In highly regulated sectors, the decision between RedHat Virtualization and Microsoft Azure Stack HCI is influenced by the specific priorities of the organization. For those prioritizing an open-source model that emphasizes transparency and multi-cloud adaptability, RedHat Virtualization is an attractive choice. Conversely, organizations that require extensive integration with Azure, efficient compliance automation, and strong hybrid functionalities will find Azure Stack HCI to be exceptionally valuable. Its direct access to Azure's suite of tools for monitoring, governance, and security makes it a frontrunner in the modernization of IT within regulated industries.

Before making a decision, organizations should carefully assess their existing IT strategies, workload needs, and compliance requirements. While both options are formidable, the ultimate choice often hinges on whether the organization prefers the open ecosystem offered by RedHat or the cohesive hybrid capabilities provided by Microsoft Azure Stack HCI.

8.6.4 For Edge and ROBO Scenarios

Hyper-V integrated with Azure Stack HCI presents an attractive solution for remote office/branch office (ROBO) and edge computing applications. Its adaptability and cost-effectiveness make it particularly suitable for organizations operating in limited physical spaces or where reducing complexity is essential. The capability to deploy with just two nodes and direct networking removes the necessity for costly high-speed switches and external storage, leading to a more efficient and budget-friendly configuration.

For instance, a multinational logistics firm overseeing remote sites in different countries can utilize Azure Stack HCI to consolidate operations and enhance productivity. By integrating Azure Arc, the organization benefits from centralized monitoring and management, which alleviates the administrative burden on IT personnel. Running workloads locally ensures low-latency performance, which is vital for real-time applications such as inventory management and sensor data processing. Azure IoT Edge further supports this by allowing containerized workloads, including AI and analytics, to operate smoothly at the edge.

Additionally, Azure Stack HCI's features, such as nested resiliency for storage, stretch clustering, and centralized management via Windows Admin Center, provide strong high-availability solutions. Its compact design and compatibility with low-core-count servers significantly lower costs per location, making it an excellent option for organizations looking to modernize their edge infrastructure while adhering to budget limitations.

These benefits, along with Azure's subscription-based pricing and hybrid cloud functionalities, equip organizations with the flexibility and scalability needed to effectively address evolving business requirements. Consequently, Hyper-V combined with Azure Stack HCI stands out as a premier choice for contemporary edge deployments, offering a harmonious blend of cost, performance, and operational efficiency.

8.6.5 For Technology Startups and Innovation-Driven Companies

Startups and companies focused on innovation often find themselves in rapidly changing environments where being agile and adaptable is essential. Proxmox and Hyper-V offer a perfect blend of cost-effectiveness and scalability, making them great choices for startups, especially in the tech or biotech sectors. These platforms allow for quick deployment and low overhead costs, enabling startups to expand their infrastructure as they develop without needing large upfront investments. For example, a tech startup might begin with Proxmox due to its open-source advantages and later shift to Azure Stack HCI as they need more hybrid cloud capabilities to enhance their customer-facing applications.

8.6.6 For Manufacturing and Industrial Operations

Manufacturing companies require systems that can swiftly adjust to changing production needs without incurring high costs. Nutanix HCI offers inherent scalability, making it a perfect fit for managing the data-heavy workloads typical in manufacturing and supply chain operations. The capability to scale horizontally by adding nodes as needed, along with edge computing solutions from Azure Stack HCI, enables these sectors to enhance production, optimize processes, and connect with IoT devices throughout their facilities. For example, a worldwide automotive manufacturer might leverage Nutanix HCI to oversee real-time analytics from their assembly lines, boosting production efficiency and minimizing downtime.

8.6.7 For Educational Institutions

RedHat and Proxmox present attractive options for universities and research organizations, especially when budgets are tight. Proxmox stands out as a fantastic open-source platform for academic settings that seek to tailor their systems to specific needs. Meanwhile, RedHat's commitment to open-source solutions and educational licensing allows universities to access high-quality virtualization without overspending. Proxmox is particularly well-suited for laboratory settings, where researchers require adaptability and oversight of their virtual environments for diverse experiments and simulations.

8.6.8 For Retail and E-Commerce

In the retail sector, it's essential to prioritize speed, reliability, and scalability, particularly during busy shopping times. Azure Stack HCI, combined with Hyper-V, is perfectly equipped to manage the fluctuating demands of retail operations. By leveraging cloud bursting features, businesses can effectively scale up during peak demand times, such as holiday sales, ensuring optimal performance while keeping costs low during slower periods. For instance, a mid-sized e-commerce business may depend on Hyper-V with Azure Stack HCI for its foundational infrastructure while utilizing Azure cloud services to accommodate spikes in traffic during special promotions or new product releases.

8.7 Hyper-V with HCI: Microsoft's Cloud-Native Powerhouse

For businesses that are well-integrated into the Microsoft ecosystem, utilizing Hyper-V alongside Azure Stack HCI is a logical progression in their IT approach. The integration of hybrid cloud solutions is becoming essential, and Microsoft's ability to connect on-premises and Azure workloads effortlessly creates a resilient IT framework for the future.

This setup is especially beneficial for companies with remote teams or those exploring multi-cloud options, as Azure Arc facilitates centralized management across various platforms. With Azure's built-in disaster recovery features, organizations can prioritize scaling their operations instead of investing in redundant infrastructure.

8.8 Proxmox: Open-Source Flexibility for Budget-Conscious Users

Proxmox stands out as an open-source solution, especially for organizations looking for affordable virtualization options without compromising on features. By utilizing KVM for virtualization and LXC containers for lighter workloads, Proxmox provides a level of flexibility that more complex systems may not offer.

For small- to medium-sized businesses, startups, and those embracing a DevOps mindset, Proxmox's capability to incorporate custom scripts, manage various workloads, and provide containerization through a single dashboard is a compelling advantage. It's a perfect fit for companies that desire the control and adaptability of open-source solutions without the hefty licensing fees associated with proprietary systems like VMware.

8.9 RedHat: Enterprise-Grade with Open-Source Innovation

RedHat Virtualization (RHV) provides a robust, enterprise-level solution tailored for businesses in regulated sectors or extensive IT settings. With RedHat's strong reputation in open-source advancements and dependable enterprise support, it has become a go-to choice for companies seeking reliability, security, and cutting-edge solutions. Additionally, by integrating OpenShift for container orchestration, RedHat broadens its capabilities into cloud-native and Kubernetes ecosystems, essential for organizations embracing DevOps and cloud-first approaches.

8.10 Nutanix: Simplifying Hyper-converged Infrastructure

Nutanix has emerged as a frontrunner in streamlining IT infrastructure through hyper-converged infrastructure (HCI) in this book. Its innovative architecture aims to remove the complexities associated with traditional three-tier systems, making it a great choice for companies that value simplicity, scalability, and cost-effectiveness. With Nutanix AHV, organizations can also save on licensing costs, which translates to significant financial benefits without compromising on performance.

For businesses in various sectors, from manufacturing to education, the ease of managing infrastructure with just a few clicks while having the ability to scale as needed brings substantial operational advantages. Nutanix's one-click upgrades and AI-driven analytics empower IT teams to enhance efficiency and minimize the administrative burdens that often come with conventional systems.

8.11 The VMware and Broadcom Equation: A Final Reflection

The acquisition of VMware by Broadcom marks a pivotal moment in the virtualization landscape. For numerous organizations, this is a clear indication to broaden their virtualization approach. The ambiguity surrounding VMware's future licensing structures and product developments under Broadcom has made many companies cautious. As we've highlighted throughout this book, depending on a single vendor can pose risks, particularly during such significant corporate shifts.

The main point is straightforward: it's time to consider your alternatives. Whether you decide to stick with VMware or explore options like Nutanix, Hyper-V, Proxmox, or RedHat, a diversified IT strategy will help ensure your organization is equipped to adapt and succeed, no matter how the market evolves.

8.12 My Final Thoughts: Building a Resilient Future

The acquisition of VMware by Broadcom marks a pivotal moment in the virtualization landscape. For numerous organizations, this is a clear indication to broaden their virtualization approach. The ambiguity surrounding VMware's future licensing structures and product developments under Broadcom has made many companies cautious. As we've highlighted throughout this book, depending on a single vendor can pose risks, particularly during such significant corporate shifts.

The main point is straightforward: it's time to consider your alternatives. Whether you decide to stick with VMware or explore options like Nutanix, Hyper-V, Proxmox, or RedHat, a diversified IT strategy will help ensure your organization is equipped to adapt and succeed, no matter how the market evolves.

8.12.1 Sustainability and Green IT Initiatives

As organizations place a greater emphasis on sustainability, this aspect is becoming increasingly important. Virtualization naturally minimizes the physical space required for data centers, but advancements from platforms like Azure Stack HCI and Nutanix are taking carbon footprint reduction to the next level. By showcasing the energy-efficient features of HCI solutions and cloud-native infrastructures, we can highlight their positive impact on the environment, particularly as ESG (environmental, social, governance) criteria become more integral to corporate strategies.

8.12.2 Innovation in AI and Automation

Another key theme that deserves more attention is the rising influence of AI and automation in managing infrastructure. Tools such as Nutanix's AI-powered analytics and Azure's machine learning for predictive monitoring and self-healing systems are essential. This focus is crucial as organizations aim to automate routine processes, enhance performance, and minimize the need for human oversight. Including a section on how these platforms utilize AI and automation for predictive maintenance and smarter infrastructure choices would enrich the discussion.

8.12.3 Vendor Lock-In and Futureproofing

It's essential to think about the potential risk of vendor lock-in. Although multi-cloud and hybrid solutions are designed to mitigate this issue, businesses can still encounter difficulties when they become heavily reliant on a single platform. It would be beneficial to highlight the significance of open standards and the value of vendor-agnostic solutions that provide future flexibility. For example, Proxmox stands out with its open-source model, while RedHat's focus on open standards helps protect against the risks of lock-in.

8.12.4 A Call to Action: Preparing for the Next Decade

To emphasize the need for future-proofing, it would be impactful to include a compelling call to action in the conclusion. Encourage organizations to not only investigate their virtualization options but also to invest in the necessary skills, training, and change management to fully harness these technologies. The technologies mentioned are more than just infrastructure; they signify a transformation in how IT departments will function and contribute to business success in the coming decade.

The choices we make today will echo throughout the next ten years, fundamentally altering how organizations function, grow, and innovate. As an author, I encourage businesses to look beyond just cost savings or short-term performance gains and adopt a more strategic mindset. It's essential to assess long-term objectives, recognize security considerations, and strive for a balance between flexibility and innovation. Whether you're looking to harness the power of next-gen AI-driven HCI or refine your hybrid cloud approach, the time to take action is now. The path to a resilient and future-ready infrastructure begins with courageous, well-informed decisions made today. Let's not just plan for tomorrow but for the entire next decade. In addition to these insights, I believe the book will delve deeply into both the technological and strategic dimensions, providing a forward-looking, actionable roadmap for decision-makers in various sectors. The concluding thoughts will equip readers to view virtualization and infrastructure modernization not merely as necessities but as a sustainable competitive edge.

The **Broadcom-VMware** episode is continuously progressing; a fresh chapter in virtualization is just starting – and it's the perfect moment for you to craft your own narrative in this exciting journey.

Index

A

Acropolis Distributed Storage Fabric (ADSF), 260

Acropolis Hypervisor (AHV), 250, 257

Acropolis operating system (AOS), 135, 249

Action plan, VMWare, 167–170

Active perpetual licenses, 86

ADSF, *see* Acropolis Distributed Storage Fabric (ADSF)

Advanced encryption standard (AES), 184

Advanced Threat Protection (ATP), 224

AES, *see* Advanced encryption standard (AES)

AHV, *see* Acropolis Hypervisor (AHV)

AI-driven HCI, 410

AI training models, 394

AKS, *see* Azure Kubernetes Service (AKS)

Amazon, 117

AOS, *see* Acropolis operating system (AOS)

Apache CloudStack, 142

Artificial intelligence (AI), 36, 129, 342, 392

ATP, *see* Advanced Threat Protection (ATP)

Avago Technologies, 14

AVD, *see* Azure Virtual Desktop (AVD)

Azure Active Directory (Azure AD), 187

Azure AD, *see* Azure Active Directory (Azure AD)

Azure Arc, 133, 406

Azure Blob Storage, 133

Azure Kubernetes Service (AKS), 158, 232, 236, 357, 386–389

Azure Security Center, 189

Azure Stack HCI, 245, 353, 358, 386, 388, 398, 405

Azure Stack HCI, use cases
  data-centric workloads, 381–386
  DR/BC, 367, 368, 370–374
  edge computing/
    ROBO, 361–366
  Kubernetes
    workloads, 386–390
  secure virtualization, 360
  VDI, 375–381

© Sumit Bhatia and Chetan Gabhane 2025
S. Bhatia and C. Gabhane, *Navigating VMware Turmoil in the Broadcom Era*,
https://doi.org/10.1007/979-8-8688-1264-4

M

GPSR Compliance
The European Union's (EU) General Product Safety Regulation (GPSR) is a set
of rules that requires consumer products to be safe and our obligations to
ensure this.

If you have any concerns about our products, you can contact us on

ProductSafety@springernature.com

In case Publisher is established outside the EU, the EU authorized
representative is:

Springer Nature Customer Service Center GmbH
Europaplatz 3
69115 Heidelberg, Germany